Creators of Gods
The Posthuman Myth

Dedicated to my granddaughter
Letizia Alamar Durán

"The way things are presented is not the way they are; and if things were the way they are presented, the whole science would be enough."

Karl Marx -Russian economist and sociologist-

Index

Pag

Prologue ......
Chapter I. The Search for Truth ......
Chapter II. A Look at the Depths ......
Chapter III. The Celestial Score ......
Chapter IV. The Alchemical Chimera ......
Chapter V. The Prophets of the Future ......
Chapter VI. First Future: Technological Generalization ......
Chapter VII. Social Dystopia: The Coercive Government ......
Chapter VIII. A New Religion ......
Chapter IX. Children of a Lesser God ......
Chapter X. A Complex Demon ......
Chapter XI. Forbidden Territories ......
Chapter XII. The Recurring Utopia ......
Chapter XIII. The City of Stars ......
Chapter XIV. The Flight of the Eternal Navigator ......
Chapter XV. An Ideal World ......
Chapter XVI. Immediate Societies ......
Chapter XVII. In the Kingdom of Heaven ......
Chapter XVIII. The New Apocalypse ......
Sources and Bibliographic References ......

Prologue

Human evolution began a process plagued by discrete leaps, inevitably driven forward, with the priority objective of preserving survival; an indispensable achievement for any animal species, of whose generic family, homo sapiens began to distance itself, decisively conditioned by the physical changes, produced in its immediate environment. In the continuous development of the road, he acquired a unique functional structure that changed drastically, drawing his future landscape. The neurobrain enabled the ability to project different scenarios, without the need for direct experimentation (Chapter II).

The totality of evolutionary changes and progress, have been faithfully recorded in their molecules, defining a portentous and perfected reproductive code, progressively endowed with greater perfection, as well as imperfect features. The genetic code currently focuses on one of the focus of cutting-edge scientific research (Chapter III). Once the dark stage defined by the metaphysical keys of magical knowledge, the formation of the first civilizations, acted as trigger of the concern about the meaning of the immediate reality, and intimately linked to this, of the meaning of one's existence. At this point, a search for answers, supported by the use of rational capacity (Chapter I), began.

The set of instincts shared with the sister species formed the basis of human knowledge, thus delineating a threshold of desires with different threshold of coherence and possibility; among the deepest ancestral beliefs, the disturbing presence of the animal feeling of survival, metamorphosed in the feeling of transcendence, concreted in the conquest of the chimera of immortality, or at least of the prolongation of existence (Chapter IV).

The promise of eternal life, previously monopolized by religious metaphysics, now coincides with the offer of commercial corporations, oriented to the economic exploitation of the applied possibilities of technoscience (Chapter V). The widespread permanence of the primal instincts and human needs have enabled the renewal of scenarios typical of a new religious fervor (Chapter X). Among the incisive effects of the newly created technomyth, comes the cystogenetic role caused by the profound and numerous changes rejected automatically, by the large conservative segment of today's society (Chapter XVII).

The modification in the optics of a new elastic and posibilist future, has built a polyhedral and changing reality, of fragile and inconsistent connotations. Global digitization will impose a first horizon of generalization of innovative technologies that will directly affect the individual and their immediate experiential environment, building the cities of the future (Chapter VI).

The irradiation effect of powerful technoscientific practice permeates both beliefs, attitudes and projects. Even notable institutions of scientific prestige, they allocate resources of all kinds with the aim of finding a new guide, this time of extraterrestrial origin that marks the guidelines of their unequivocal path to the security of the future, forgiving the assumption of one's own and non-transferable civilisational responsibility (Chapter XI). Another of the broad social groups, perfectly delineated and socially distinguishable, progressives and advanced, contemplate the imminent possibility of exploration and conquest of new planets, avoiding in parallel, the increasing risks arising of a planet wounded and attacked by the extractive mentality prevailing in the capitalist economic system (Chapter XIII), launching an inca saber search for new stellar habitats (Chapter XIV).

The constant desire for improvement and the achievement of utopia, however distant it is drawn on the horizon, will push those individuals eager for experimentation and overcoming their real or perceived limitations, to the maximum use of cyber options (chapter VIII), as well as by extension to their direct descendants (Chapter XII).

Science on its unstoppable path of instrumentalization of the discoveries obtained will represent, as has traditionally been, a dual role in the implementation of technical achievements. On the one hand, it will provide an explanatory view of the reality that each time acquires a greater degree of complexity (Chapter IX), all the more so, the greater its spectrum of application. Unfortunately, the desiderátum of a perfect world is not feasible (Chapter XV). The alternative focuses on the search for an organized, sensitive and thoughtful and ethically superior society (Chapter XVI). Only in this way is there an authentic possibility, both for the abandonment of false expectations based on false and illusory promises, and of overcoming the hyperboles, dystopian and apocalyptic visions, launched by the nefarious prophets of the future, that do not abandon the attempt to fly the traditional messianic discourse, as well as the exercise of the banners proper to the pretentious utopia and forming a complete eschatology of the future (Chapter XVIII).

During the book's exhibition, seemingly similar argumentative criticisms are presented, but it actually responds to assertions made by different groups or authors and, above all, of varying degrees of significance or scope. We hope to have achieved a pleasant reading, while underpinning proven and scientifically demonstrative facts.

Valencia, October 2019.

Chapter I. The Search for Truth

"No one who is curious is foolish.
People who don't ask questions
remain ignorant for the rest of their lives"

Neil DeGrasse Tyson - American astrophysicist and communicator -

"Science is the great antidote to poison
of enthusiasm and superstition."

Adam Smith - British Economist -

The way in which human beings have perceived and explained the reality of the world around them, has gone through two distinct phases, although overlapped even today. The first of these corresponds to the mythical or religious knowledge, also called magical or superstitious knowledge. This primitive and equivocal way of interpreting natural reality, has facilitated the emergence of mysticisms and practices whose most noticeable result, can undoubtedly be classified as harmful and negative, not only for past societies since remains today, causing biological, economic and serious damage to the health of unsuspecting believers. Included here is a long list of practices and purported pseudosciences, such as shamanism, sorcery, witchcraft, healing, vision, tarotism, divination and many others. The alternative and subsequent form of explanation of reality, was born with philosophy, anteroom of the sciences and was developed with the advent of Greek culture.

Greek civilization originated in Cretan and Mycene culture. He received in parallel Turkish influences, due to the Doric and Ionian invasions mainly. It occupied the same geographical territory as it is today, located in the southeast of Europe, with direct access to the Mediterranean Sea. He received in full the ancient traditions, based on the mythical account, with the undisputed seal provided by cosmological beliefs, typical of the ancient polytheistic religions that directly affected and therefore delayed, the progress scientists from both the Greeks and the Romans themselves; for example, Hipatia of Alexandria, found its main problems with the understanding of the Universe, thinking that as perfect structures, both the orbits and their original forms, should be pure spheres; until the time of accepting ellipsis as an alternative way in the configuration of planetary systems, it had to be a long time. A direct proof of the influence of polytheistic beliefs lies in its architecture. The Partheno, the best-known monument of Greek culture, was erected for the worship of the goddess Artemis. Later, however, the second major Greek construction, the Agora, contained illustrative meanings of drastic ideological changes. The fundamental purpose of the completely flat form of construction of the construction was to protect citizens from inclement weather. Its interior structure, divided into several types of rooms, were intended to create suitable places for dialogue and exchange of ideas.

The transformation of the perception of reality, introduced by the Greeks, would end up constituting in the classical stage of their culture, the nuclear mode interpreting the real phenomenon: the worldview with the greatest influence in the future of the world, in practically the all areas of thought. While observing the celestial vault, the first substantial turn from the past, was not to seek the answers in the celestial composition, but rather to inqueathe the right questions that needed to be answered; ¿What is nature, what is the point? And above all, ¿what is the role of man and its meaning? And if the answer to the latter question is yes, what is its end? Until that time, the cosmos had constituted the great oracle for ancient cultures, such as the Maya or the Aztec; the stellar configurations had been erected in the indispensable predictor in the forecast of the crops, in their annual calendar, in the director of their destinies, and above all, represented the mansion of their multiple gods.

The second essential component, in the new vision of reality, rested on the method of questioning, in the face of uncertainty and complexity in the observed physical reality. Departing from the futility of mythical arguments, the new form of questioning of reality focused the mechanism of response to essential questioning of reason, previously endorsed by direct observation. This almost perfect binomial, formed by observation and reason, is at the base of the upcoming development and universal implementation of the general method of science, which would constitute the elementary pillar of scientific progress; not in vain, the first Greek philosophers were called naturalist physicists.

Probably influenced by the discoveries they were acquiring, or simply by the insulting magnificence of the observed nature, they conceived an identification that even today, maintains a certain influence on scientific thinking: the identification between beauty and truth, which has inherently remained in the direction of scientific search. Even further, if truth was necessarily beautiful, virtue as a basic component of knowledge should preside over man's behavior. In this way, the true cultural revolution of humanity began, classified as the second evolution and whose effects, have exponentially accelerated human evolution. Indeed, the phylo-sophie is the historical beginning of the search for truth in human development. A second contribution, no less decisive, was the repeated finding that the supposed and desired perfection of nature and therefore of physical reality was non-existent.

The only possible existing apotheosis was in the abstraction of that same reality, that is, in the idea. In uttering this principle, Plato, the dualistic philosophical current, still in force at the present time. Greek civilization has meant the conceptual eruption of greater historical relevance, in terms of the sowing of questionable seeds by structurally forging, drastically and unanswerably, the present Western thought. At that time in history, philosophy and science, they obtained their official birth certificate. The philosophy intended in its beginning, the knowledge of the essence of every entity visible from an analytical perspective, rational and unlike the bias dominant in the future sciences, with a primary approach of a holist character, that is, from a global prism and Generalist. The research method rested on reason -the logos-. From that moment on, reality would be explained through philosophy and future sciences; art through myth and fantasy, and religion, through faith. Thus, the whole of lived reality used as explanatory instruments of reality, simultaneously, reason, myth and faith. The admiration for the whole is at the origin of philosophy -called physis- that sets its focus, in the immediate environment or cosmos.

Early philosophers are described accordingly as naturalists and cosmologists. One of the determining coordinates, in this way of focusing rational attention, on two different entities, artificially separated and concrete, in the immediate physical environment and the cosmos, leads to the definitive establishment in rational thinking, of the duality, a similarly present rule of law, in natural evolution; cause and effect, well and bad, day and night, constituted a way of interpreting the facts, forming the basis of linear and cyclical thought.

The exterior different from the self; the finding of the difference between the self and the outside. A new idea emerged here, both in the practical and intellectual sphere, that will gradually take body in human thought and action, to the point of becoming one of the concepts-strength of political activity; the distinction and progressive departure of the individual from the community, continuing the line established by the first contrived distinction between individual and nature. Assuming the risk of generalist simplification, philosophy as the governing body of knowledge and above all, the attitude inherent in its own development, set the path, of two major areas of action that delineate today, the situation of society so-called postmodern. Scientific knowledge and political philosophy and its embomition, in models for the design of societies. A clear example is starring Solon, one of the so-called 7 sages of Greece, who lived around 600 BC and who expounded his political thinking about the government of the city of Athens.

From the reading of this elegy, eunmia and dysnomia -good and bad governance- emerge as opposing figures, in the manner of personifications of those qualities, which the cops must adopt or reject. The epic poetry –Homer, Hesiod– had mentioned both terms, but it is the lyrical poet, who seeks to witness his performance, as a legislator with the fervor of his verses. Later, Plato and Aristotle, will present in their political writings, issues related to the dichotomy represented by "good and bad government".

The contribution to contemporary politics of the early Greeks does not end here. Today, fundamental concepts, developed for political discursive action, are used. The incorporation of the philosophical concept of the Greek ethos will become the current ethic. It is simply a question of the application of moral standards to the relationship with others, to the whole of the community. At a later time, it will transmute into a heading of permanent debate between the different ideological tendencies, which persists to this day.

The Greek polys or city, contemplates the birth of the state, with the congenital purpose of service to the citizen; this is both its end and its raison deem and reason for being. The person who is engaged in the administration and management of the city and therefore, of the citizen collective, has the ultimate objective of ensuring and preserving the general interest, the well-being of the social whole. This dominant and practically unique feeling governs the thought referring to the outside of oneself, to the ethos. With regard to the second field of development, the distinction between the public good and the well-being of the population is a relatively recent intellectual construction.

For the Greeks, the public sphere of action and discourse meant, in the face of the need to live together, assuming the irreducible plurality of the human condition. The policy was completely separated from the work of producing the goods necessary for survival; wholly oblivious to the usefulness. Its fundamental task was aimed at the collective achievement of the good life -eudaimonia- but not in the sense of application of scientific principles such as those of medicine, but rather, the highest objective set in obtaining happiness or Welfare. Political activity was composed of a set of activities that they were to bring the being closer to excellence -aristeia- and immortality; contributed to improving citizens by establishing a reciprocal relationship between personal and public virtue.

Either in its Athenoid version, as a balance between the two planes, or in the Spartan modality, of total submission of the personal plane to the public. In any case, the split between ethics and politics, was maintained until the arrival of Christianity; until that moment, there could be no one without the other. Contributions, in the theoretical area of politics as human experience, can be categorized as profound and varied in Greek civilization. Zeno introduced dialectics and Georgias rhetoric, both used in subsequent political activity, albeit as mere discursive strategies. Greek thought, travels the vast territory of human curiosity producing germ ideas, conditioning the future in its full extent. A clear example is the moment, when Democritus introduced the idea of atomism into future physical science.

The sophists shifted the old general spotlight from the cosmos to man, automatically introducing the concept of areté or virtue, that is, the appropriate form of inner behavior in the relationship with others; Hippocrates established the premises of modern medicine, with the introduction of a diagnosis followed by personalized treatment. In this period, the two largest figures of Greek thought and both produce contributions of universal projection. Aristotle introduces the distinction between individual and collective problems, that is, between ethics and politics. This dichotomy has dramatically damaged political activity throughout history. It also makes another equally damaging contribution, this time, in the systematization of knowledge through taxonomy or classification of the general body of knowledge in specialties, subsequently completed by Descartes.

The influence of Greek thought on religion is equally significant and decisive. The last vital period of Greek philosophy corresponds to the emergence of Hellenistic schools of thought: cynoism, epicureism, stoicism, skepticism and eclecticism. It coincides temporarily, with the beginning of another of man's future practices, imperialism, embodied in its Greek version in the figure of Alexander the Great who was instructed by Aristotle himself and advised by his father, built the greatest empire in history, achieving the union of three of the continents of the earth.

The basic conquering desire, existing in imperialism conducive on one side, the imposition of a particular cultural system and another, the transmission of a concrete culture that is coercively implanted in the conquered territories and peoples. Thus, Alexander the Great personalizes the initiative of union of mycene culture and the Persian people. This in turn, during the reign of Cyrus, spreads within its broad limits, the standardized currency. Another of the historical parameters of that time, focuses on the migration of civilization from East to West. It is definitively noted that the emergence of great civilizations such as Egypt, Mesopotamia, China, Iran and Phoenicia and consequently the facts determining the human future, produced within each of them.

The Sumerian culture of Babylon brought the pictorial writing, discovered in the city of Kish, in Iran, 3,500 years BC. The emergence of writing allowed the accumulation of knowledge, an alternative to the oral tradition that existed until that time and the only channel of transmission, experiences and experiences of human groups. Writing is the starting point of culture, considered strictly. Medical science was born within the philosophical mindset. Other series of notorious contaminations such as ancient etiological rationalism, becomes a conceptual priority of medicine today and forms the basis of the medical model, guided by the simple "cause-and-effect" scheme. In another order of facts, the Phoenician people, the architect of one of the most remarkable commercial epics in history, in their eagerness to promote trade between civilizations and due to the need to establish a system of understanding with them, invented and contributed to the flow the alphabet, which would later be incorporated by the Greek language and serve as the basis for Latin, as well as the rest of the known languages, and in particular to the languages of Indo-European root.

The classification or taxonomy of knowledge, while favouring further disciplinary scientific development, has significantly undermined the unitary understanding of scientific knowledge. Aristotle established the dissection of human problems into three categories, in force today implicitly. Problems or physical issues, logical and ontological problems –cosmogonic– and thirdly, introduced the discussion on moral issues. On the other hand, Plato established in his work "The Republic", the forms of government possible in the new reality, before which man stood: the great collective. A concentration of individuals who formed a community; an absolutely new phenomenon for humanity. Once discarded by ineffective, autarky and anarchy or absence of government, he also established, the transformations of legitimate modes of government, at the time they are corrupted or deformed. This outlining has proved basic, for further discussion in the design of social models throughout the story. In essence, the community can be governed by three kinds of distinct figures.

The first represents the exercise of power through a single individual. In ancient history, the one-person regency constitutes the most widespread form of community leadership and is represented by the king. It maintains a direct correspondence with the prehistoric model of leadership, exercised by the strongest or wisest individual of the herd or tribe. Secondly, the Community leadership may be the responsibility of a small group of subjects who, because of their knowledge, experience and knowledge, assume the direction of the collective. This congregation, called the counsel of sages and its prehistoric appearance, also corresponds to an evolutionary stage of the formation of power in the tribal clan. Lastly, the power of the community rests with all citizens, manifesting itself as the power of the people.

Democracy, which through direct participation in its beginnings, determined the direction of the city or community. All these forms of address, in their correct application format, were considered legitimate. In this scenario, the generation of laws was considered a pre-political activity; necessary for the existence of the city-state, just as the walls were. For the state to be viable, the laws were needed; for this reason, legislative activity was detached from productive work -poiesis- and craft activity, being configured as an activity of the wealthy class.

The history of Greco-Roman philosophy extended to 529 AD, when Emperor Justinian closed pagan schools, whose knowledge and tradition will be recovered and grouped by Arabs in full middle age. At this point, the official implementation of religion as a substitute for knowledge occurs by political decision and therefore becomes possible in the new reality, before which man stood: the great collective. A concentration of individuals who formed a community; an absolutely new phenomenon for humanity. Once discarded by ineffective, autarky and anarchy or absence of government, he also established, the transformations of legitimate modes of government, at the time they are corrupted or deformed. This outlining has proved basic, for further discussion in the design of social models throughout the story. In essence, the community can be governed by three kinds of distinct figures. The first represents the exercise of power through a single individual. In ancient history, the one-person regency constitutes the most widespread form of community leadership and is represented by the king. It maintains a direct correspondence with the prehistoric model of leadership, exercised by the strongest or wisest individual of the herd or tribe.

Secondly, the Community leadership may be the responsibility of a small group of subjects who, because of their knowledge, experience and knowledge, assume the direction of the collective. This congregation, called the counsel of sages and its prehistoric appearance, also corresponds to an evolutionary stage of the formation of power in the tribal clan. Lastly, the power of the community rests with all citizens, manifesting itself as the power of the people. Democracy, which through direct participation in its beginnings, determined the direction of the city or community. All these forms of address, in their correct application format, were considered legitimate. In this scenario, the generation of laws was considered a pre-political activity; necessary for the existence of the city-state, just as the walls were.

For the state to be viable, the laws were needed; for this reason, legislative activity was detached from productive work -poiesis- and craft activity, being configured as an activity of the wealthy class. The history of Greco-Roman philosophy extended to 529 AD, when Emperor Justinian closed pagan schools, whose knowledge and tradition will be recovered and grouped by Arabs in full middle age. At this point, the official implementation of religion as a substitute for knowledge occurs by political decision and therefore becomes in the model of knowledge prevailing in the cultural sphere. It is ultimately the official continuation of the historical trend, initiated by the Abrahamic tradition.

In this way, medieval political thought Christianized the heritage of Roman law. Governing consisted of conducting public conduct, in accordance with the law, and power was considered sovereignty. In it, the prince or king represented without discussion, the world's dominating agent. By completing the circle, human law was contingent upon divine law, but maintaining its essence; governing therefore consisted in banning; the law should not be applied, but rather established, and as in divine law, the sovereign possessed the maximum legal precept: to deprive or grant life. By extension of this principle, citizens' possessions, such as sustenance, housing, property and other fundamental rights, became the subject of government attention and even worse, were transformed into public goods.

During medieval times, for Christian doctrine man is defined incapable of accessing virtue by himself; precise for this, inexorably of divine grace. A budget is installed, which has endured to this day; the government of the state or earthly city cannot aspire in any way to form virtuous citizens. As an extension of this principle, the common good in its tomist formulation -according to St. Thomas- is a custodial of the exercise of sovereignty governed by civil law, a transfer and direct transcription of divine law.

Thus, in the Middle Ages, the common good achieved full identification with justice; their compliance by individuals, as legal subjects, would attain virtue, yes and only if they were considered good compliance with the fair laws. In the modern era, thanks to the legacy preceptal bases, governments of totalitarian identity, both individualistic and state,, apply this same principle. For all these governments, all individuals and citizens become a population. This is how biopolitics arises. The account inherent in governments becomes a pastoral discourse and public welfare, it is seen as a specific attribute of modernity. The welfare state, at the same time, represents a guiding concept and an instrument of power and has not been able to be built outside religious ideology. Regarding the encompassing trend typical of ancient philosophy, we had to wait until the 17th century, to observe a scientific trend of totally opposite sense. Physical science has led to this strong trend.

The primary pretence was aimed at reducing the obvious complexity of reality, through the use of mathematical formulas. The formulation should explain the natural behavior, presided over by the conditions of simplicity and beauty. There have been many and remarkable examples in this direction from Newton to Einstein, with a special mention, which many consider the most beautiful formula in the history of physics, attributed to Paul Dirac, prestigious French mathematician. The equation was presented in the late 1920s, allowing the connection of two separate scientific disciplines so far, such as quantum mechanics -which describes the behavior of very small objects- and the behavior of objects governed by the laws of electromagnetism.

Certainly, and regardless of the degree of universality in their application, the formulas developed by illustrious physicists and mathematics have shown a clear inability to conceptually explain reality, outside its own scope. In this sense, the emergence of two theories of encompassing intentionality, such as the information theory, proposed by Claude Shannon and Warren Wiever and the theory of systems, whose initial proposal corresponds to Von Bertanfly, are recognized by the bulk of epistemologists and science theorists, as the theories of greater potential with respect to its practical breadth, as well as its application flexibility.

The concept of system brings together components and approaches, coming from the physical sphere, such as the field theory proposed by Kurt Lewin; chaos theory -whose butterfly effect, is widely known thanks to the graphic image described in the phrase "the movement of a butterfly's wings in a given place, can trigger a hurricane anywhere else in the world" and the laws of the Thermodynamics. In this way, a return to the observation of reality from a broader view takes place, to some extent similar to the constitutive of the primitive philosophy. It is known as "system theory", a set of interdisciplinary contributions, which include the study of the defining characteristics of systems, that is, of entities made up of interrelated and interdependent components.

The German biologist Karl Ludwig von Bertalanffy, proposed in 1928, his "general system theory" as a broad tool that could be shared by many different sciences. This approach contributed to the emergence of a new scientific paradigm, based on the interrelationship between the main components of any type of system. Previously, systems as a whole were considered to be equal to the sum of their parts and could be studied from the individual analysis of their components; Bertalanffy, questioned such beliefs. The fundamental contribution of this approach lies in the simple concept of interaction between the components of a systemic whole. Since the creation of the general theory of systems, it has been applied to biology, psychology, mathematics, computer science, economics, sociology, politics and other exact and social sciences, especially in the context of analysis interactions.

Today, the concept of complexity has progressively mutated to the point of shaping the so-called sciences of complexity. The leading institution in this field of study is the Santa Fe Institute located in California. The characteristic of the institution's theoretical research and production shares the general intention of today's generic science. It is concrete in the application of its explanatory principles, to all fields of scientific research, both in its basic and applied modalities. From biology to space navigation, to agriculture or city design. At present, these fields remain the priority focus of today's science.

The first, in a way from smaller to the largest physical size, concerns the behaviour and nature of the essential components of matter. The atom and subatomic particles are studied thanks to expensive particle accelerators. The only one currently existing, CERN is located in Switzerland and has 37 kms. length. The research technique used focuses on a single strategy; by accelerating at a speed close to light, technicians cause the shock of atomic particles, in order to analyze, their decomposition into smaller subparticles, with the ultimate intention of finding the minimum particle, the fundamental principle in the generation of matter that makes up matter.

The second object of study is focused on the human brain. In this sense, nothing has changed in recent decades, regarding the great mysteries and areas investigated. Using techniques based on magnetic resonance imaging and electroencephalography, specialists analyze a wide variety of fields of interest, such as memory, emotion physiology and brain activity, generated by different stimuli of reality. Alongside neurochemical research, the study of the human genome attracts the efforts of today's science, based on the discovery of the DNA double helix, made by Watson and Crick, in 1953.

Observation of the universe, mainly through astronomical tubes and observatories, unites two objectives of interest considered essential. On the one hand and as a target of greater recognition, the existence of planets outside the solar system itself or exoplanets is sought, in order to determine their possible future habitability by humans. On the other hand, attempts to discover the origin and laws of formation of the universe continue, among which the analysis of background radiation, or primeval radio signal, seems to stand out among all others. In this same vein, the attempt in space observation continues, focused on the discovery of radio signals of an artificial character, distinctive of the existence of extraterrestrial civilizations. Some critics, highlight in this activity, the human need of ancestral origin, consisting of finding guides or paths, marked by entities considered superior, whether of alien religious origin.

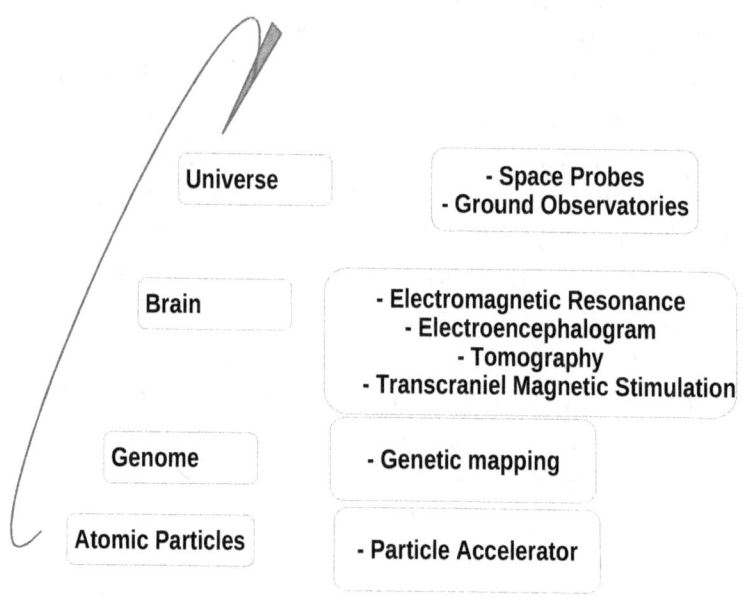

The great fields of basic science research have undergone virtually no substantive changes in recent decades. In contrast, the techniques applied are increasing in quantity and precision in parallel to technological progress, facilitating the results obtained. Source: Own elaboration

If defined areas of study are further observed, it can be perceived as the particle accelerator, pursues the discovery of the elemental particle that formed the universe, while astronomical observatories and spatial probes, they in turn attempt to discern their origins, so their objectives fully coincide. On the other hand, the study of the human genome, contemplates as one of its main lines of research, deepening the evolutionary changes that ended in the complex structure of the modern neurobrain; convergence means a single study objective. Thus, when science theorists claim that at present, the two major fields of research focus on the study of the questions offered by the universe and the brain, it is a completely true assertion. The main sources of questioning of today's science, fully coincide with the original attempts of ancient philosophy; the permanent questioning of the nature of man and the one fabric that surpasses him in magnificence and power of attraction: the universe from which he comes.

Finally, science in general operates by using two conceptual methods essentially: models and theories. The theory involves a higher level of abstraction than the model and is responsible for providing the theoretical explanation -causes and consequences- of any discovery or result. The model, on the other hand, has a practical and operational character. Determines the method, instruments and techniques used in a particular investigation. For example, in memory research, the model is considered a heuristic, that is, a specific instrument, such as the illumination of nerve connections by laser beams. Subsequently, the theory will be responsible for explaining the different brain areas involved, as well as the connection between them and the distinction between short- and long-term memory. This general operation addresses the study and research ofThis set of scientific actions, have marked the human course to the present, in which the aforementioned state of well-being, is strongly marked by medical care for citizens, with the wide range of possibilities that such a fact entails. In this way, two types of well-being in the population are clearly configured; public or punctual welfare, other than that offered by private initiative, aimed mainly at the economically middle and upper classes. Thus, in the present century it is undebatable that science and technology, now collectively called technoscience, are considered the most responsible meanings for human well-being.

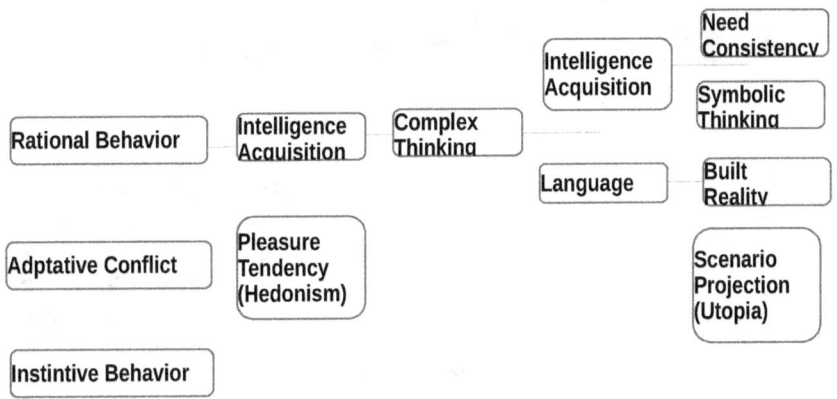

Nature operates through the principle of parsimony. In the case of human evolution from an upper mammal, the chimpanzee experienced constant changes led b the inmediate physical environment, increased the prior brain structure of the primate and capating it to achieve the upright displacement to thtough the ingestion of animal proteins up to a certain structural size that enable a number of fuctions and potencialities never before existing including abstract thinking and language development. The evolutionary process incorporated other complementary changes in body physiology such as those produced in the larynx, pelvis and lower and upper extremities. Source: Own elaboration.

Chapter II. A Look at the Depths

> "Science is made of data,
> like a house of stones.
> But a lot of data is not science
> more than a lot of
> stones is a house."
>
> Henri Poincaré -French Mathematician-

It has been known since the middle of the last century that in the human evolutionary process, the hominid separated from the chimpanzee, only about 6 million years ago. Despite the false belief that a large part of the human genetic burden is made up of insignificant DNA, sometimes referred to as "junk DNA," it is obvious that if those genes that turn out to be the specific agents and dynamizers could be identified of human evolution, the "molecular clock" responsible for acquiring the distinctive features of homo sapiens could be accelerated or improved.

In this line of reasoning, it has been discovered that only 15 million of the more than three billion base pairs or "letters", which make up the human genome, separate it from chimpanzees; only 1.5% of total DNA differs from the two species, chimpanzees and humans. Precisely at this point, it should be emphasized how the passage of mammal superior to homo sapiens occurred and this mode is none other, than the overlap of the typically human intelligent behavior, to the primary intrinsic structure of the chimpanzee.

The specific process of hominid evolution began 6 million years ago, with the genetic separation of chimpanzees and thus, 1.4 million years ago, the first bipedals emerged. They lived in the forests, until the great Rift fault -a geological structure that extends from north to southern Africa- separated the African continent into two halves; the western part separated from the eastern part, varying in that process the climatic diversity that enable disputing the obtaining of essential resources for life, totally different in both halves. In the eastern cone, which extended to the Mediterranean Sea, the large collision became in the Savannah as a predominant habitat, unlike the western part, which retained the original tropical climate, with abundant vegetation and rich food resources for the herbivores and consequently, also for their predators.

The glaciation resulting from the great continental collision, was permanently installed in the eastern strip of the continent, causing the disappearance of at least 20 species of prehominids. At an age not exceeding 2 million years, the first humanoid to walk upright, the homo ergaster, reached a height of 1.70 cms., with a cranial capacity around 900 cc.

-cubic centimeters. New structural changes emerged in his body physiology, facilitating the acquisition of the final upright position. The morphological structure of knees and feet, underwent appreciable modifications; the preprehensibability of the hand previously with functionalities typical of arboreal life, was extended and flattened. But without a doubt, the substantial modification took place in the pelvic area, a real nucleus in the construction of the current human skeleton. The pelvis became a narrower, yet more rounded piece, thus facilitating delivery and surrounding itself with a more specific muscle configuration. The spine acquired an arched shape, with modifications at the bone and muscle level, centered on the abductor and buttock muscles. By far the substantial modification occurred at the point of connection between the spine and the brain, protecting itself with greater coverage offered by the muscle clustering of the neck.

Another of the decisive modifications took place in the hands, with the strengthened appearance of the oponicability of the thumb, an indispensable premise for the consolidation of the prehensile function, allowing the subtle handling of objects that ultimately enabled the manipulation of the stone in the early stages of the Paleolithic period and, above all, in the Neolithic period. The prehensil capacity was improving, to the point of allowing instrumentality, basic condition for the manufacture of tools, hunting instruments and weapons; poly-instrumentality subsequently arose, which led to various added practices, such as the dissection of prey for feeding and making warm clothing. This polyvariety of tasks, drove the sapiens away from his ancestor ape for good.

The adoption and subsequent consolidation of the upright position of hominids, combined with the need for food, led to a nomadic way of life, facilitated by the requirement of monitoring animal species, usual prisoners, indispensable for her feeding and dressing. Catapulted by improved mastery of instruments, and acquired mode of displacement, the distances traveled were extended to the point of exceeding the geographical boundaries of their natural habitat of life. The hominids, and more specifically, homo habilis, starred in the great human migration from the African cradle to the entire globe.

Prior to the start of the great migration, two of the first hominid species shared the same precarious environment. Drought and cooling defined the immediate environment of both species. In West Tanzania, parantropus was hunter-gatherer, feeding on tree roots as main nutrients, data demonstrated by the dental records of this species showing a flat configuration, more suitable for chewing Vegetables. Although they possessed certain already human characteristics -such as laughter- their general features were closer to their direct ancestors. While homo ergaster, he was engaged in scavengerry, looking for pieces of leftover or abandoned meat. Ingestion of the meat protein –long-chain lipid foods– in addition to contributing to overall body development, fundamentally determined the increase in brain volume, which at that time reached 900 cc. The process would continue until reaching 1450 cc., typical of homo sapiens and giving rise to a complex neurobrain.

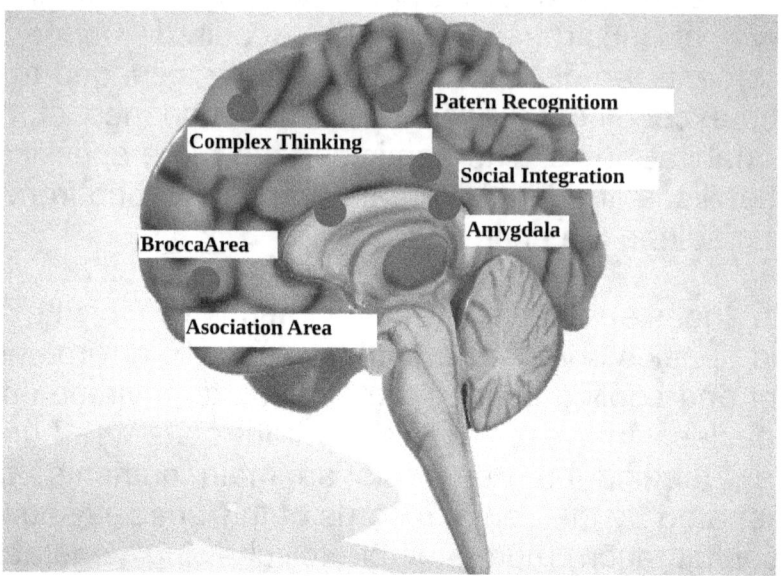

The human brain reflects asan exact code its development by the different evolutionary phases from its beginning as a reptil through the greater structure common to mammals and from that moment on the forward growth of homo sapiens characterized by the area of association in the prefrontal cortex base of the abstract thinking and above all, the use of language.
Source: Own elaboration.

The brain in its first phase came from reptiles, with little performance except for purely basic ones, related to food, reproduction and survival. In the evolutionary transition to higher mammals, fundamental areas for herd and later conviviality are developed in the tribe, such as the capacity for social interaction. Following the evolution of upright displacement, the development of the ability to recognize patterns became indispensable for the realization of the clan's fundamental tasks, such as hunting or recognition of the hierarchy - the varying degrees of authority, which represented the possibility of social survival as a whole. The essential nucleus of the evolutionary process is located in the amygdala, center of the instinctive behavior, responsible for perceptions of fear or anger, decisive in survival behavior; the exact place where Descartes would later locate the human soul.

Finally, already in the last part of the exclusive brain development of humans, three specific areas were incorporated, located in the prefrontal cortex. The areas of abstract thinking and association processes have defined the physiological basis for progress and social evolution. They stand out as the third joint functional area, the Brocca area and the center of Wernicke, -with indispensable connections with the auditory center and memory-, physiological centers of the complex language.

The functional connections while-brain areas, represent simultaneously, the basis of high intelligence performance, as well as, the main barrier to its study and decryption. Katherine Pollard of the University of California has designed a computer program, with the aim of defining the areas in which the rapid changes had occurred, responsible for the specifically human evolution. He used a giant computer located at the University of California, Santa Cruz. The results obtained were positive, both quantitatively and in terms of their relevance. It found 201 regions of the human genome, showing accelerated variation. Among them, a succession of 118 bases that together have been known as accelerated human region 1 -HAR1-. But in addition to this, of the total set formed by these bases only 18 mutations have meant significant alterations, from the moment humans emerged.

The next step in research focused on the attempted decryption of the precise characteristics and the specific functioning of the newly discovered genetic concentration. The main conclusion stated that this area had shown significant stability, during millions of years of evolution. Primates separated from birds about 300 million years ago, yet only two base pairs differ between chimpanzees and hens. Thus, the HAR1 zone had remained virtually unchanged for several hundred million years, with only two changes in the letters of genetic composition, G and C -the hereditary chain contains 4 basic components identified by its initials, the Adenine, Cytosine, Guanine and Thymine- and yet, in just six million years, the HAR1 area had mutated 18 times, a fact that could explain the enormous acceleration of human evolution. The fascination of the research team increased with the discovery of the role that the HAR1 area had played in establishing the general arrangement of the cerebral cortex famous for its wrinkled appearance, triggering the increasing intelligence Human.

The substantial characteristics of human intelligence are based on symbolic thinking, for which the development of language has constituted a true catalyst, accelerating and perfecting its development. Homo sapiens possessed the morphological characteristics, necessary to develop an increasingly varied and rich language. The management of articulated language has led to the possibility of the development of large human groups, the beginning of the first civilizations and in parallel, the cultural evolution responsible for the so-called second major evolution or evolution cultural, whose distinctive characteristic rests on the fact that its development, has followed an exponential rhythm, that is, not linear and continuous, but rather represented by a curve that expresses an acceleration of logarithmic type.

Located in the Brocca area, in the temporal lobe of the brain, is the language area. According to studies at the University of London, the agent responsible for this difference with the chimpanzee is located in human chromosome number 7 and has been called FOXP2. Using the DNA analysis technique in the search for mutations -considered the key element of the evolutionary process- it was surprisingly discovered that all modern humans come from a common ancestor. The "mother" of all humans has been christened Lucy mitochondrial, settling her figure permanently in the scientific panorama of paleontology –he figure of the biblical Eve has its scientific counterpart-. The antiquity of the mitochondrial mother, goes back 300,000 years, even though once the genetic filum of homo sapiens has developed, identical phenomenon -the appearance of a common cellular mitochondria- was repeated again, only about 60,000 years ago. In short, all the current humans belonging to the entire race that populate the world, are descendants of a single woman.

The role of the mutation has been sufficiently described as the enabler of evolution, however, its random event condition has been poorly highlighted, i.e. due to an error in copying the genetic code. This is, however, the main key responsible for population changes. Nature also makes mistakes in its continuous process of change and has used error, as an element of progress and above all, has been decisive in generating such a relevant trait, as increasing the diversity of populations. In addition to the first evolutionary region described in genetic topography, another area called HAR2, seems to be responsible for fine digital skill, that is, enabler of finger dexterity, necessary for the manipulation of delicate tools. The crucial gene for brain capacity development, called ASPM.

A team of scientists from the University of Edinburgh isolated the particle RIM-941, the only one discovered whose existence appears only in homo sapiens and not in any other primate species. In addition, geneticists can show that this gene appeared between one and six million years ago, that is, after humans and chimpanzees separated. This particle is responsible for a crucial fact in human evolution and behavior, because at every image, the brain automatically associates an emotion, such as happiness, fear, anger or envy, and this process occurs in the brain amygdala -an area located deep in the temporal lobes of all upper mammals-. The connection between the two brain hemispheres -right and left -provides the feeling of a consciousness of its own, of being a unified self, from the existence of this hereditary reagent of a unique nature. The relevance of this discovery confirms the already known process, concerning mental functioning, which through the categorization of external stimuli, constitutes the functional basis for the formation of memories and on this plane, proceeds to the building the individual consciousness.

This resource is the starting point, in the elaboration of hypotheses and projections, made around the meaning of the human being and its nature. Spirituality, religion or cognitive psychology and constructivism, concentrate at this starting point all kinds of elucubrations, theories and generalizations. Undoubtedly, it is one of the discoveries of greatest significance in the whole of scientific, cultural and social activity, due to the profound implications of all kinds that can be derived from its study and analysis. In fact, there are indications that human evolution on a large scale, that is, the basic form of the human body and the level of intelligence, has practically stopped. For this reason, the continuation of the evolution of the species aimed at an increase and improvement in its capabilities and potentialities, currently depends on specialized research and more specifically, the applied modality of research. Yet there are still unknowns to clear and speculation, focusing mainly on the human genome. For example, ¿if it contains approximately twenty-three thousand genes, how is it able to establish a mechanism of control over the connections between the hundred billion neurons, which total one billion connections? -one followed by fifteen zeros-. It seems mathematically impossible. The human genome is too small, it should have a volume about a trillion times its size, to be able to encode all neural connections. Thus, human existence itself is very close to an impossible mathematician.

The answer to this question could lie in nature's own mode of operation. In particular, it seems to use chance to create a complex structure of a level comparable to the neurobrain, although this assessment results only from a superficial observation. First, a large number of neurons seem to establish their connections -synapses- seemingly randomly. This fact, if true, would rule out the study option used by brain topography, which seeks to build a detailed map of brain mass. The intellectual resource of teleological tone or finalist, sometimes called intelligent design, charges in such events all its strength. The comprehensive key is the detailed observation of nature's way of action and not projective analysis of its possible finalist motives. Such a purpose is non-existent, since the randomness is not really such.

The brain connections are organized progressively after the birth of the individual and progress permanently, interacting with the environment or physical environment, throughout the existence of any person. Nature uses modules or parts that are repeated continuously. Once its functionality or utility has been tested, the process is repeated constantly. The very origin of the cosmos, seems to use this formula in the construction of stable molecules. This fact could explain the prominence of few genetic changes in human evolution, characterized by the spectacular growth of intelligence, noted in the last six million years.

This way of operating nature, is known as the principle of parsimony and represents the characteristic, leisurely and constant mode of progress. However, some specialists are not tempted to question the enormous adaptive potentials, studied in children between the ages of 0 and 6. Physiologically, they present a frankly incredible range of potential development. Starting with the possibility of resisting water ingestion when submerged, due to the high position of the larynx that does not descend until the beginning of language acquisition, around the age of 2 years. From that time until the age of 6, its ease of language learning, is classified as potentially endless. Children who attend multilingual schools learn the different languages naturally as well as the mother tongue. It is in this time frame, that the brain shows a greater number of brain connections.

This fact clearly indicates and expressed in simple terms that children at this age are potentially able to perform any type of future activity. The volume of neurons and their connections, declines significantly during the process of specialization of the activities of the subject. In other words, the individual's specialized choices determine the number of nerve connections needed and the disappearance of unnecessary brain resources. The process of nerve development that begins at the time of fertilization continues its accelerated growth after birth. Interaction with external stimuli facilitates certain nerve brain connections and inhibits others, while myelin -a substance that progressively coats nerve structures- covers the new electrical bonds that are established Gradually.

One of the perceptual processes experienced by all children up to the age of two, consists of synoesthesia and only 2 % of the adult population can experience it. This phenomenon is defined as the feeling of fully combined experiences and is due to the extraordinary connections between different nerve centers. It occurs in the cingulated center, placed in the upper area of the hippocampus -responsible for memories-. The surprising result consists of an almost arbitrary mixture in the interpretation of the various external stimuli. Thus, a child in those ages can savor the voices, see colors when caressed or perceive music when they observe a toy.

This pleasant and suggestive dysfunction, in adulthood is mainly among artists or creatives. Returning to the great milestones of evolutionary dynamics, the emergence of the great civilizations enabled by the massive social order, constitutes one of the basic axes on which human civilization is based. The codified language is responsible for this phenomenon, which constitutes the third great singularity found in the historical perspective, coupled with the cosmological exceptionality -the formation of the universe -and the biological particularity- the appearance and development of intelligent life. The sum of the three types of irregularity produces the defining key of the current moment of civilization. Leading social theorists have studied phenomenon of great human societies. Among them, the following include J.R. Habermas and Niklas Luhmann. Both attribute to complex language, the functional possibility corresponding to social order. The organization of large societies -which in animal life, does not usually exceed 125 members- can reach millions of individuals, under social norms and codes of high degree of complexity. The spoken tradition played a necessary, yet unique, role for the transmission of knowledge, but it has been writing, the solid pillar on which to build scientific knowledge.

The common language is a communication code. From here, another series of codes such as logic and mathematics, vehicles of expression normally associated with the different degrees of human knowledge, were developed, forming a historical pairing. The first level of knowledge is based around the descriptive level at which the common language is normally used. Secondly, the explanatory stage allows knowledge of the causes and characteristics of a given phenomenon. You can add to common language, formal logic, and math. The third cognitive step, rests on the possibility of prediction of the event or phenomenon studied. Finally, the reproduction of the investigated case constitutes the fourth rung of knowledge. Here it is incorporated, the programming code or algorithmic. The latter represents the degree of knowledge that today's civilization possesses, in a large number of disciplines and scientific activities. Thanks to this latest type of code, the programming of simulations in their different modalities, whether based on autonomous agents - reproduces the role of different relevant social agents - or genetic algorithms - uses the rules of genetic transmission to apply them to social norms- allows to visualize virtually any kind of real or virtual situation that is desired, enabling a degree of understanding and understanding that only offers the perception of an image. In addition to this, the brain produces a series of substances aimed at prioritizing the performance of the subject during and after, the realization of survival efforts and also consequent, the strengthening of social relations. In the first case, mainly the muscle system receives an extra dose of energy. To do this, the stimulation of the adrenal glands that provide adrenaline to the blood system is triggered. The autonomic muscle and nervous systems, acquire increased abilities translated into greater reflexes and muscle strength, shaping a set of behaviors, well known and labeled as survival syndrome. The brain also produces various substances beneficial to the overall functioning of the body.

Among them, dopamine, immediately consistent with performing physical exercise -this is the reason for the feeling of well-being that is felt when performing a sporting activity- as well as ingestion of pleasant foods. Likewise, oxytocin is present in situations of sexual intercourse and infatuation. Similar function possess esorphins, also called "happiness hormones", used to combat stress and despondency situations, as well as phenylalanine with similar performance. All this, due in principle to a difference of 1.5% in the genetic configuration that distinguishes the human being from the chimpanzee.

Chapter III. The Celestial Score

> "Science will always be a quest,
> never a real discovery.
> It's a journey never an arrival.
>
> "Karl Raimund Popper -Austrian philosopher and professor.-

Today's technology is made up of a series of scientific specialties, which act in an interconnected way and with a common objective. Among them, bio-engineering and within it, engineering and gene editing clearly polarize scientific activity. If the old but continuous demand for multidisciplinary among the different sciences, has been repeated incessantly by critics and scientists, today, a large number of research teams, distributed in universities and research, public and private around the world, research from different fields, identical problems Since its discovery, by Watson and Crick in 1953 the human genome, genomics seems to have become the great hope for racial improvement and evolution of humanity. As if it were a musical score, belonging to one of the great musical geniuses of history, it is investigated with increasing interest, constituting a true epicenter of ideas, performances and proposals, some of them absolutely utopian.

The ENCODE Project represents the most significant of the joint efforts, aimed at deciphering and cataloging, the functioning of the components of the human genetic code that can provide relevant units of information. Funded by the Institute of American Health, it began its journey understood as a pilot phase in 2003. Three years later, 1% of the total genetic map, selected on the basis of criteria of biological interest, had been analyzed. In the first phase, the organizational bases, technologies to be used and analysis systems were established. The partial results, together with the part considered central, were published in successive articles by one of the most important scientific journals such as Nature.

The second phase was covered between 2007 and 2011, publishing its results in 2012. Without going into technical details, in this period we studied central aspects in the role of proteins in RNA -Ribonucleic Acid or Messenger Acid- and mapping of the transcription and modification sites of DNA -Deoxyribonucleic Acid-. This central chapter was accompanied by 20 additional projects in which more than 400 specialists worked, belonging to 30 research institutes, distributed throughout the globe. One of the most outstanding results for today's bio-medicine was the discovery that certain specific combinations of genes and certain sequences of genes, ordered by the RNA, show a close relationship with certain diseases, including cancer. Efforts are focused on the decryption of these protein assembly processes and also those processes in which they have not yet been observed.

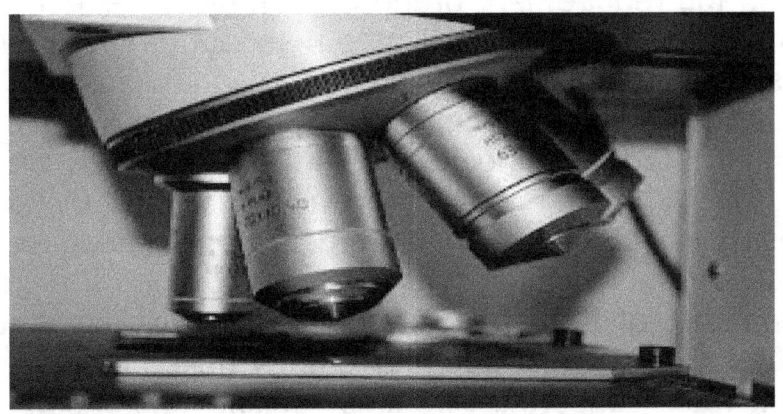

The optical microscope has been one of the observation techniques used for the observation of disease producing agents such a viruses and bacteria and also the basic components of biological beings sucha as human cells in their different varieties. One of its varieties, the optical microscope has allowed different research techniques such as optogenetics in which the neural connections on wich to operate or investigate are iluminated. Source; Pixabay.com

One of the most important aspects of global research is based on the requirement of the initial project, sequencing the human genome and entails the mandate of the funding institute, on public access to discoveries and findings of the investigation. It has produced not only results, but has built models that concrete the regions -which make up up to 70% of the human genome - can play a relevant functional role, as parts of the sophisticated system of control of biological organisms Upper. It is obvious to note, the great media impact of the project and its influence on global biopolitics, that is, in its potential use by governments and large private consortia. Genetic editing increased its potential for action, when in 2005, Dr Karl Deisseroth of the University of Stanford brought a new technique to genetic research called optogenetics. It consists of the combination of flashes of light, either from a laser or an LED, to some relevant group of neurons that encode proteins and at the same time, are sensitive to lighting, called opsins. This technique allows to identify populations of cells, involved in different diseases of genetic origin, facilitating the observation of their behavior and consequently, the search for new treatments for their healing. In 2010, the journal "Nature Methods" distinguished this research instrument, as the most important of the year.

Potential applications range from preventing epileptic seizures to sleep manipulation and similarly to studying more common processes, such as hunger or addictions and drug dependence, as well as Parkinson's disease. In 2009, Dr. Jennifer Doudna, Professor of Chemistry and Molecular Biology at Berkeley, discovered and introduced a new tool for genetic editing called Cas9, small in size and capable of becoming a true genetic scalpel of great precision, aimed at the removal and pasting of tiny parts of the genetic chain. A technique called CRISPR -Short Grouped and Regularly Spaced Palindromic Repetitions, in Spanish translation. Subsequently, the biochemist Feng Zhang, led by a team of researchers from the Harvard Institute and the Massachusetts Institute of Technology, provided a description of the mechanism for coordinating changes experienced in the cell genome, both mice as human beings. The acronym CRISPR was coined by the renowned Spanish scientist, Francisco J. Martínez Mojica.

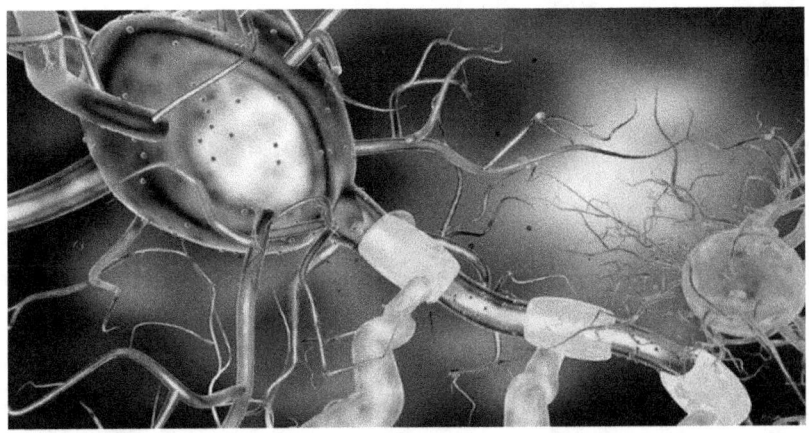

The lighting of nerve connections through different techniques use light-emiting diodes (LED) similar to those in mobile phonesthat allows to highlight the relevant cellular connections studied. Source: Pixabay.com

Recently, Chinese scientist He Jiankui claimed to have created the first genetically modified embryos, using the CRISPR/Cas9 technique. An action aimed at increasing resistance to infection, caused by the HIV virus - acquired immunodeficiency syndrome. Research in gene therapy progressively finds greater fields of application, extending to areas with little commercial appeal, such as rare diseases of hereditary origin. The consequences of this type of genetic surgery have greater implications than those that focus simply on the repair of individual altered functions, since by operating at the genomic level, they are transmitted hereditaryly to the descendants of the treated individual, thus preventing the reproduction of the anomaly repaired in their descendants. Thus, it is seen from an overview of population therapy.

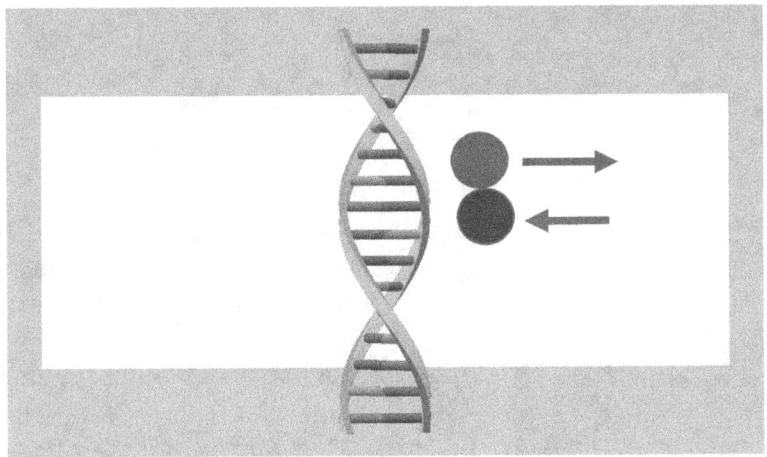

The CRISP technique allows as if it were a biological scalpel, to removed damage parts of cell and replace them with a completely healthy identical component. In this way, some potentially harmful genetic alterations are discarded achieving the disappearance of diseases, malformations and dysfunctions of genetic origin. In addition, to the extent that its operates with the same gene, the changes will be transmitted through inheritance to the descendants, thus achieving the definitive establishment of the correct genetic transmission. Source: Own elaboration.

Chapter IV. The Alchemical Chimera

> "The idea of death, the fear of dying chases that creature called a human being like nothing else."
>
> **Ernest Becker -American anthropologist-**
>
> "We should call science only to all the formulas that succeed Always. All the rest is literature."
>
> **Paul Valery -French writer and philosopher-**

Since its inception, the human being has sought an explanation of the nature around him and closely linked to this, to the meaning of his own life. The first relevant finding on this point lies in the perception of his own conscience and the firm conviction of being one in front of the environment and also, one different from the others. His sense of relative dominance over habitat combined with the subjective perception of impotence in the face of nature. He then began to develop unconscious relativism.

It was superior to the situations that occurred on Earth, but it felt inferior to the tremendous phenomena that normally came from the heavens; and for that reason, he began to look up at the stars, a behavior that has not stopped repeating for millions of years. He also deduced, that as it happened in its natural habitat, all fact had a responsible cause for his generation. Therefore, there must be beyond heaven, some higher being capable of working those magnificent events that sometimes benefited him, how in the case of rain and in other situations it caused considerable damage, as in the case of storms, the cold, the rays and the death itself or that of their loved ones. Thus, primitive animism -providing life with an inanimate object, a representation or symbol- and innate fear, came together to shape the ancestral religious feeling, which against the opinion of some anthropologists and sociologists, anchors the roots of its beginning in the fear, as a predominant feeling. It must have been at least two million years before I found a coherent explanation for the complicated world and its laws. The stars and the gods, one for every force of nature, possessed their own enormous, almost limitless powers.

Likewise, the promise of eternity was consolidated during a historical period, in the widespread belief that the known universe responded to the ultimate expression of divine creation. In this context, the Creator should not encounter any greater problems, to configure other space-time dimensions, in which to locate places of eternal rest; heaven and hell. In this way, a recursive loop is established, established by both religion and philosophy. In both cases, the physical laws are violated in their entirety. This philosophical conception is inexorably linked to the essence of time. The arrow of time, as perceived by the human being, always flowing forward constitutes a cosmic necessity, as well as a perceptual demand; In other words, there is no possibility of returning the egg yolk to its original position, within the reconstructed shell, while not being found, the visit of inhabitants of the future or of the past. The time travel, so used and admired by sci-fi lovers, is cosmologically unworkable.

The doctrines that explained the world also provided the formula for obtaining one of his most distant desires; if he complied with certain laws, he would live forever, for ever and ever; in essence, it would achieve eternal life. The Christian religion in particular assumes a cyclical eternity or looping existence. Reincarnation violates all general physical principles and in particular space-temporary principles. A substantial nuance must be distinguished in this matter; immortality has a different meaning to the meaning of eternal life. The latter responds with complete fidelity to the religious promise made to Viking, Muslim and Christian warriors. All monotheistic religions make the promise of eternal life their greatest offering and without exception, as a reward or reward for certain actions, both in a situation of peace and in war.

Religion has historically been the greatest motivating axis for belligerent behavior and conquest; and not precisely of gifts or virtues, but above all, of lands and riches; not to save souls, but to exterminate enemies; not for the benefit of the poor and needy, but for the excessive increase in power and profit of the coffers of the rich and powerful.

The philosophical doctrine of dualism acted as a catalyst for belief in eternal life. The fundamental motive of the persevering human belief in the concept of eternity rests on its coincidence with the usual perception of reality. In most cases, reality is categorized as the coexistence of opposing phenomena, applied to both ideal and physical planes; day and night, good and bad, man and woman and in Eastern philosophy, yin and yang. The German philosopher Friedrich Nietsche also defended the idea of the ancient return. Ludwig Wittgenstein offers an idea of eternity, in the sense of breadth and absence of limitation. Likewise Martin Heidegger, provides a vision of time, based on the false premise of temporal understanding under the religious budget of eternity. Thomas Huxley coined the term agnostic, defining the conviction of disbelief, first of all that which is not demonstrable and objective in the year 1869.

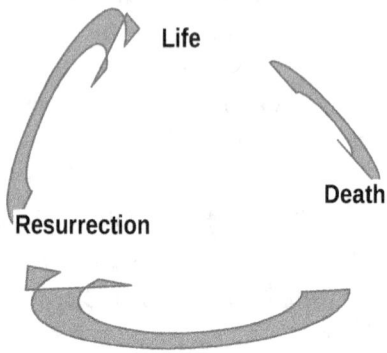

The time loop consists of a cyclic repetition of events. Its is a widespread belief of episodes and the same temporal course used in history, philosophy and also physics. Source: Own elaboration.

Existentialism as a doctrine of life, concrete dwality in permanent doubt and continuous questioning about the meaning of human existence, has produced opposite currents, such as Christian existentialism as opposed to atheist. Both Soren Kierkegaard and Jean Paul Sartre, certainly the two most recognized authors of this doctrine of thought, both allude to atheist existentialism. Belief in eternal life, however, is not generated by religious doctrine and is rather pre-it. In the 1960s, a group of paleontologists discovered in Japan's Yamagata province, mummification techniques in which body organs had begun to dry out before death. With the aim of transcendence of the physical world, to obtain more years in paradise and to attain the previous enlightenment of reincarnation, the Buddhist monks of Shington, mummified as a symbolic act, oriented to achieve the salvation of humanity, thus preceding the heroic act of the crucified Christ that narrates the New Testament.

In this practice, the monk would only eat fruits and nuts for three years, then pine needles and tree bark, for another three. He would even drink poisonous tea, to keep insects away from his body. Finally, the ascetic would meditate on a grave using a breathing tube. By ringing a bell occasionally, he signaled that he was still alive; when the chimes ceased, their acolytes scelded his grave. After three years, the tombstone would open. If the monk's body had no decomposition, he was worshiped as a living Buddha. Even if he failed, he was buried with special honors. Of the many hundreds, only 24 monks have attained the status of "living Buddha".

Eastern religions, such as Buddhism and Hinduism, include in their doctrine metempsychosis or reincarnation, which professes the possibility of progressive perfection through successive reincarnations. The final release is to stop the cyclic wheel, caused by an individual ego, which it considers a kind of being totally illusory. Likewise, the concept of cyclical time was already found in Egyptian culture; could occur in or out of time. In philosophy, both defenders and opponents can be found in the incorruptible nature of the soul. Aristotle believed that matter, movement, and time were eternal. However, the legend of Prometheus and Sisyphus advocate the denial of eternity.

One of the most influential scientists of all time, he lived in a world in which science overlapped with superstition and pseudoscience. Before chemistry really existed as a discipline, Newton practiced alchemy and the hidden arts, even holding the belief that messages within the Bible predicted that the world would end around 2060. More than a decade of Newton's reflections were lost when his lab burned down, but the rescued texts suggest that he also devoted himself to the search for the Philosopher's Stone. He had amassed numerous books on this issue, including studies by Flamel, another well-known alchemist. He worked on these projects in secret, as the British Crown pursued alchemists. While experimenting with alchemy, in the second half of his life, he suffered a nervous breakdown. With a tendency to insomnia, apathy, loss of appetite and paranoia, probably all due to mercury, arsenic or lead poisoning intake. Many scholars believe that he was actually successful in achieving important alchemical discoveries, but that he destroyed all evidence before his death.

Alchemy constitutes pseudoscience, which has sought with greater emphasis to make the ancient myths related to immortality a reality. The elixir of eternal life, the holy grail that promised eternal life to those who drank water poured within and, of course, the Philosopher's Stone. Sometimes revered, sometimes persecuted; some mystics, some phonys; alchemists were accused of heresies, spells and frauds, but they were undoubtedly the inventors of well-known laboratory instruments and also developed research methods used in today's science.

However, its fundamental goal was not the application of formulas aimed at achieving the transmutation of metals, nor did it even lie in the discovery of the elixir of eternal youth, but was mainly oriented, to the search for a pure, perfect and orderly universe. Alchemy dealt with the subtle forces existing in nature, as well as the various conditions and states of matter. This obsessive practice involved a secret, esoteric and initiative combination of certain knowledge drawn from the natural sciences, empiricism, metallurgy and natural philosophy. A secret combination, in that his adepts hid their knowledge from the laypeople; esoteric principles, because their fundamental principles did not derive from scientific theories; and iniciates, because only one who was guided by a teacher entered into it.

For alchemists there was only actually a single material –the raw material–, presented, in multiple forms and in different states. Depending on the proportion in the combination of these elements, the different minerals and therefore metals emerged. The latter, they had a life of their own and evolved; in addition, all of them –except gold and to a lesser extent silver– were considered impure. According to his beliefs, no living being, nor any inanimate substance, was perfect; not even man had managed to complete his physical perfection –in that he suffered from insurmountable diseases– nor spiritual –they firmly believed that man was dominated by evil–.

The efforts of alchemy, focused on trying to transform vile metals, into the idyllic state of gold and in general, in the pursuit of maximum perfection in any aspect of their lives. Through the Alchemical process, called the Great Work, matter was manipulated to achieve its transit through a series of states: fire, being heated; air when distilled; water, being cooled; and land, by precipitating, while its purification was achieved to end up becoming the "raw material" or uncontaminated element, from which all that has been created comes.

When the alchemist obtained this element, he continued the process until he obtained the philosopher's stone, sum of all perfections and transmulator of metals; a few grains, turned into gold any vile metal with which they came into contact- and in the form of dissolution, achieved the longed-for metamorphosis in the elixir of eternal youth, also called universal panacea, curator of all evils. The Philosopher's Stone would get the perfect man, by ending his physical imperfections – freeing him from all sickness and granting him immortality- and spiritual —if he had reached the goal, it was because he possessed universal knowledge- and would lead him from return to the happiness of Paradise.

For the initiate, the obtaining of the philosopher's stone and also the eternal elixir, it meant not only the consequence of a process, but above all, constituted the reliable proof of the conquest of the parcel of truth, to which he had dedicated his life and that, consequently, he had personally achieved happiness. Perhaps the first manifestations of alchemy can be located in Alexandria, where the pre-alchemical practices of greek, Chaldean, Egyptian and Jewish cultures were collected and updated. But many centuries before, on the other side of the world, in China, alchemist practices have been dated – with the main objective of making gold. The Orientals considered that by ingesting gold, a man could athaeuntious powers and above all, immortality. Moreover, because gold proved difficult to find in nature, Chinese alchemists tried to manufacture it artificially. In the 10th century, the Arabs began practicing alchemist techniques, based on the ancient Chinese and Greek methods, which they knew through Syria, Persia and Egypt; to them are the concepts of elixir of eternal youth and philosopher's stone and among their contributions to science, highlight the discovery of ammonia salt, the preparation of caustic alkalis, the discovery of the properties of animal substances and the introduction of the distillation decomposition method for the analysis of these substances. Its classification of minerals formed the basis of most taxonomy systems and also of the current periodic table, which would later be used in the West.

They were also responsible for the coinage, between the 8th and 9th centuries, of the al-kimiya voice –alchemy - on the basis of the Greek word chymeia , an art of fusion of metals, which in turn comes from the ancient name of Egypt, Chem. The fact that many alchemists also practiced magic, astrology and Kabbalah led to the suspicion and accusation of the practice of black magic and witchcraft. Although many nobles – including kings – and clergy – including popes – practiced discipline, the various states and above all, the Church pursued for many centuries their supporters, judging and condemning some to the bonfire, sometimes more for defending er heretical theories, than for their alchemical practices.

In Rome at the end of the 3rd century, Emperor Diocletian, had the treaties of alchemy destroyed and his followers executed in Egypt. During the Middle Ages, they were accused of magic, all those who experimented with physics or chemistry. The Archbishop of Prague was persecuted, on the charge of alchemist by the Council of Constance (1414-18) and a decree proclaimed in Venice in 1530, prohibited such practices under the death penalty. It is not surprising, then, that in this context alchemists used a symbolic, secret and hermetic language, to which only adhered to adheres to the adheres initiated by a master and equally, who avoided their persecution by signing their works under false name and attributing their thesis to illustrious authors – Alberto Magno, Raimundo Lulio or St. Thomas. Even circumstantially, they titled their works, with concepts taken from the Holy Scriptures, in order to dye their theories with the appearance of respectability and Christian feeling, thus avoiding the dangerous accusations of heresy. But as is obvious, together with these scientific alchemists, many of farcemakers emerged with great power of conviction, claiming to be in possession of the secret of the transmutation of any metal into gold. His claim lay in obtaining gold from the bags of the unwary.

Perhaps, the most famous of all alchemists was Teophrastus Bombastus von Hohenheim (1493-1541), better known as Paracelso –remembering his wisdom that of the famous Roman doctor Celso–, who has gone down in history, among other reasons, for founding the so-called yatrochemistry, an excision of alchemy that studied the medicinal use of all types of substances. Paracelso, made these substances from mineral acids, metallic salts, alkalies and all kinds of plants, called arcane. His doctrine was based on the belief that the cause of any disease was caused by the excess or organic defect of a particular substance; consequently, those active substances, contained in the plants, would have the power to re-balance the organism, in addition to feeding and purifying it. Thus, Paracelso erected a barrier between alchemy and science, laying the foundations of a chemistry based on exact measurements, in which he went so that he developed processes aimed at isolating, mixing and dosing the substances, which were always part of a combination, with the pretension in their medical activity, to obtain mineral and plant extracts with the highest possible dose of purity.

Despite its lights and shadows, alchemy has been considered the origin of current chemistry. Not for nothing, their practitioners invented stills, furnaces, flasks, beakers, filters, special glass containers and many other utensils and devised procedures and working methods, such as sublimation, distillation, evaporation, coagulation, fusion and calcination, which continue to be used in today's scientific laboratories. Although at the time they were branded mad and visionary, today's science has given some credit to their essential beliefs. His theories, about the possibility of transmutation of metals, the unique origin of matter and all creatures, and the atomic composition of all substances of nature, although rudimentary, are now irrefutable. For example, thanks to today's particle accelerators, the transformation of mercury into gold is already possible, albeit at an unsustainable cost and requiring so much energy volume, which is unfeasible.

In the same field, more than half of the world's platinum production goes to car catalytic converters, suffering a logical increase. Recently, scientists at Penn State University used a laser to obtain the decomposition of an electron from a tungsten carbide molecule, giving it the properties of platinum. Tungsten carbide -a metallic compound that costs one thousandth of what platinum costs- is not a noble metal, but it will contribute to the fall in the price of platinum. It is expected to successfully repeat experimentation with groups of tungsten carbide molecules, with the aim of finding substitutes for other rare elements. Like alchemists, this research group has not yet been able to find the method that makes it possible to convert lead into gold, but it is enough for them to imitate the rest of the elements of the periodic table.

At the same time, apart from these points of encounter with the "official" science, the activity of the alchemists, has remained during the time surrounded by a halo of mystery, which clouded its true historical role. Fake or illusory; madory or dreamers, the truth is that alchemists –the true – only pursued perfection, full knowledge and immortality, three of the dreams or chimeras common to all human beings through the times.

The new prophets of technoscience, make identical promises with the old objectives, by manipulating the beliefs of the world's population, previously sensitized by the use of mass media.

The Universe is now conceived as a four-dimensional space with three spatial dimensions and atemporal one, except for excepctions such that offered by string theory. Any choice of existence outside these dimensions as that preached by monotheistic religions is considered cosmologically inviable. Source: Own elaboration.

Basic scientific research has provided a rational and measured response to the anxious and compulsive human claim, encrypted in the attainment of eternal existence. New technologies catalyze the fiery fire, in which this widespread and ancestral hope is macerated. At the National Cancer Research Center (CNIO), dozens of birds and mammals have been studied and come to an interesting conclusion: there is a very clear relationship between the lifespan of each species and the shortening rate of its telomeres, some biological structures located inside cells, intended for the protection of genes, located on chromosomes and already known, play a substantive role in the aging process. The relationship discovered is susceptible to mathematical expression, by means of an equation capable of accurately predicting the longevity of the species in question. The formula shows a positive correlation between the two variables: the rapid shortening of telomeres and the lifespan of an organism; In other words, those species whose telomeres are shortened, live less time.

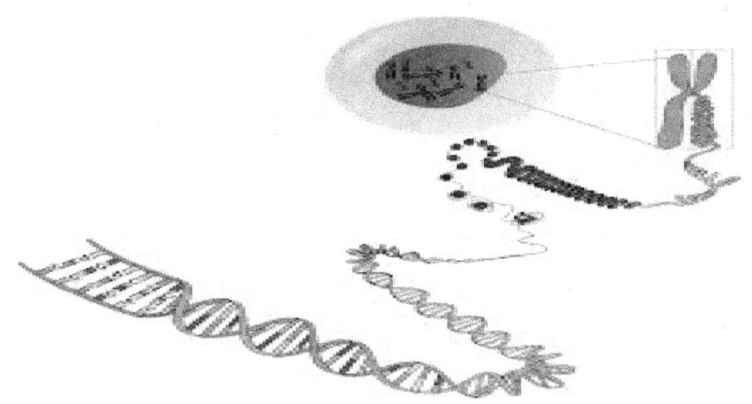

Telomeres form the ends of chromosomes intended for cell protection and responsible for cell reproduction through a self-copyng process (right part of the image). The telomeres are shortened to a time when they are unable to produce new copies.

Dr. Maria Blasco, head of the research group, explained in the scientific journal "Proceedings of the National Academy of Sciences" (PNAS), the type of mathematical equation that the formula represents. The relationship conforms to a particular type of mathematical curve, known as a potential curve that is also observed in the study of other processes, such as population growth, city size, species extinction, body mass or income Individual. This type of curve, represents the prototypical form of complex phenomena and is well known by experts in this study specialty. For this reason, it fits perfectly with the objective data relating to the half-life and maximum lifespan of the different species; in the human case, 79 years on average and the 122 years achieved by a French citizen -Jeanne Calment-.

Telomeres, integrating the ends of chromosomes, inside cells and their main function, is the protection of genes. However, each time the cells multiply to repair their own damage, their telomeres are shortened. In this process there is a time, when telomeres do not have the ability to regenerate again or get too shortened. Normally when that happens, the cell stops working, i.e. it dies. This is the true biological barrier to limiting the life-long life of species; the alchemical promise, the pursuit of eternal life, thus finds a biological obstacle of an insurmountable nature.

Until recently, the relationships between cell telomer length and lifespan had not produced positive correlations. There are species with very long telomeres that live little, and the opposite occurs. Human telomeres, on average, lose about 70 pairs of bases -the bricks of genetic material- a year, while the telomeres of the mice, about 7000 pairs of bases.

In studies conducted in collaboration with the veterinary team at the Zoo in Madrid, blood samples have been taken from dolphins and elephants. The results indicate that the rate of shortening of telomeres predicts the longevity of species much better than other parameters considered so far, such as body weight -in general smaller species tend to live less time- or the rhythm heart.

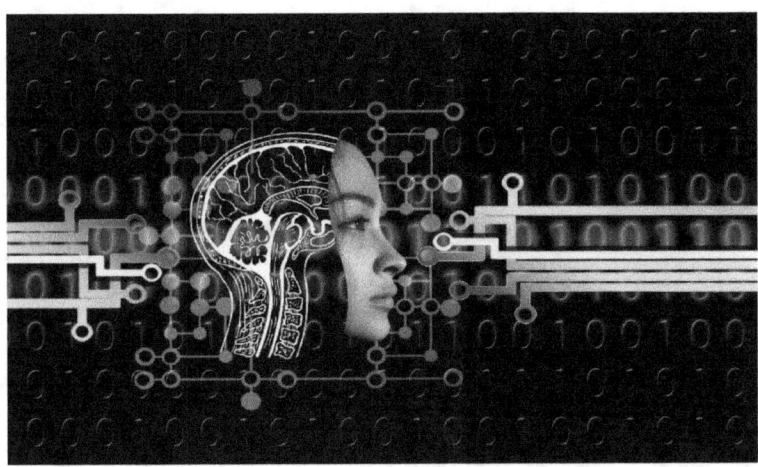

The greatest promise made by the new prophets of the future is oriented to the possibility of achieving a life of virtually unlimited duration. To do this, human memories would be transferred to cybernetic support. It would actually be a vegetative existence, lacking the possibility of experiencing emotions that take place in the human body's own neurochemical system, modeled through millions of years of biological evolution. Source: Pixabay.com

Chapter V. The Prophets of the Future

> "I am one of those who think that the
> science has great beauty.
> A scientist in your lab
> he's not just a technician: he's also
> a child placed in the face of phenomena
> naturals that impress him
> like a fairy tale."
>
> Marie Curie -French Scientist-

> "The most useful science is that
> which is more communicable"
>
> Leonardo da Vinci - Renaissance Scholar-

Saul of Tars, after his conversion to Christianity, called Saint Paul, did not personally know Jesus of Nazareth, since his first stay in Jerusalem, took place at least 30 years after the death of the Nazarene. However, he became the ideologue and chief architect of Christian doctrine, due to a supposed appearance of Jesus, already after his crucifixion.

He was the main diffuser of Christianity, which at the time, was not yet an independent religion. He used the means of disclosure at his disposal. He wrote epistles to the fledgling Christian communities, including the Galatians, Romans, Philippians, Corinthians, Thessalonians, Ephesians and Colonenses and relevant figures of the moment such as Philemon, Timothy and Titus. His travels spanned the known cult world, Rome, Crete, Judea, Jerusalem in particular, Syria, Damascus, Galatia, Turkey and his own hometown, Tarsus in Cilicia. In current terminology, he would have been the CEO -chief executive of the company, in its Spanish translation- of Christianity.

Its main objective was twofold. On the one hand, to highlight the figure of Christ, like the messiah announced in the Old Testament, since a new doctrine on a central figure that would provoke a permanent attitude of waiting, on the part of his followers and secondly, the transmission of the new belief. His task greatly facilitated the pointed and detailed explicitation of Christian doctrine, which took place in the Council of Nicea in 375 A.D. The institutionalization of the figure of the Messiah, would change over time in the personalization of the leader, in principle of a religious nature, but subsequently, widespread to other social fields, mainly political and social.

During the summer of 1956, some scientists met at a Congress in Darmouth, New Hampshire, USA, with the aim of discussing how machines were building, which were behaving seemingly intelligently. In particular, it was based on the conjecture that every aspect of learning or any other characteristic of human intelligence could be described so precisely that a machine would be capable of incorporating a computer program, intended for the simulation of a particular characteristic. Some participants, such as Allen Newel and Herbert Simon, preferred the term "complex information processing," which literally fits the computer process, but John Mc Carthy, the event organizer, insisted that the name". Artificial Intelligence", would attract the attention of the general public and attract many articles about the new field of research; And indeed it was.

Mc Carthy partnered with Marvin Minsky, to found the Massachusetts Institute of Technology (MIT), the Artificial Intelligence Laboratory where applications have been developed around the fledgling discipline, for more than half a century. Neither is considered the creator of the study area, a fact that is fairly attributed to the British Alan Turing, but it did delimit the starting point and generalization of matter. It represented a revolutionary starting point, which has accelerated the technological development of the 21st century.

Thus today, applications have been developed incorporated into the area of personal use, both as well as work. These include cloud computing -the cloud- big data, blockchains -secure exchange systems- along with bitcoin, virtual reality, the Internet of Things, different types of robots, autonomous vehicles and dark factories -so called, since there are no human operators and no lighting is needed-, 3D printing, synthetic biology, molecular self-assembly, or new and hopeful quantum and organic computing, among others. Not to mention the huge literary and film-production of science fiction, conducted around AI -Artificial Intelligence-.

Both Stanley Kubrick and Michael Chrichton came to this laboratory to consult with both founders, details on their masterpieces: Odyssey 2001 and Jurassic Park, respectively. Raymond Kurtzweil earned his bachelor's degree in Computer Science and Literature from the Massachusetts Institute of Technology (MIT) and was a student of Minsky. He is currently one of the founders of the University of Singularity, sponsored by NASA and Google. At the same time, it has been incorporated by this same corporation -Google- as a special advisor in the area of technoscience.

He has become an undisputed leader and guru of the technological future of contemporary society. Millions of people have attended their talks and presentations, either in person or through broadcasts through plasma devices or any other digital medium. The footprint originated by the term "artificial intelligence" in the social collective has been impactfull on a global level. It is probably one of the most successful cases of trade name or branding of recent times. Indeed, few brands have achieved a positioning -a place in the consumer's mind -beyond the one that their products have achieved.

Perhaps in this category, it can be placed to Coca-Cola which began its commercial journey in 1886 and expanded all over the world. Its brand and design are already part of the cultural history of the twentieth and twenty-first century. Another successful example is represented by Apple's "the bitten apple" logo, about which millions of its fans propose curious interpretations about its meaning, from a reference to Adam and Eve, to a tribute to Alan Turing, who died at the bite an apple poisoned with cyanide. Likewise, the icon of the Nike sports brand -a line with volume, with a degree of curvature that suggests an upward trend- has been the subject of psychopedagogical descriptions of origin that we sincerely fail to understand in its entirety.

It is noteworthy how many official reports or scientific publications on AI -artificial intelligence- even in scientific articles reveal a great lack of rigour in the use of some concepts or technical expressions, especially by using not to consider the differential meaning of a given term, depending on the context of the area of knowledge in which it is handled. Thus, expressions of marked tautological meaning and generalist application are used, which reduce or pervert the proper meaning of the particular term, whatever it is. However, even considering artificial intelligence and especially robotics as an essential component of the great promises of today's technoscience, its implications result in a greater and also deeper spectrum of influence.

First, the typical model of American marketing, aimed at the construction of some of the modern myths or generally accepted realities, such as the cyclical existence of economic crises, has been used, has been affirmed that economic cycles, manifest the shape of a wave with their peaks and soils, in a cycle around 7 years of cadence-. A certain trade name or impact mark is clothed by a mantle of supra-cultural relevance, through the use of prestigious and socially valued elements, such as scientific articles, news, documentaries and reports monographic, present in the great mass media, especially those of visual content, such as television, news networks and many websites of recognized influence.

Second, once the overall impact on the desired sociological debate has been obtained, the dominance of the new trend becomes controlled by large corporations, albeit with the necessary individual prominence of the charismatic leader, in this case, Ray Kurtzweil, but not in a unique way. Calicó is a corporation owned by Google, specializing in studies on aging. The aforementioned University of Singularity (Singularity University) assumes the undisputed prominence in the promotion of the technological future. Promises about the goodness of postmodern society seem to have no limits. Claims such as "the first human who will live 1000 years has already been born" or "aging is simply reduced to a technical problem", emerge fluidly from this modern dream factory, contradicting or simply eradicating, the most biological rules Essential. In the same vein, the Institute for the Future of Humanity has been definitively established in this field of work, under the patronage of the University of Oxford. It also adds to the defense of the transhumanist movement, the Transhumanist World Association -Humanity +- led by Nick Bostrom and David Pearce.

It has artificialized, in the American operating mode and with the power emanating from the influence of Anglo-Saxon cultural imperialism, the ultimate business of the future thanks to marketing and promotion strategies inherent in commercial marketing Classic. Today, for example, any average citizen can request with a simple saliva sample, a complete analysis of their individual genome. The cost is only figured, in two months of waiting and 1000 dollars. In short, we are witnessing the discovery, implementation and launch, of the definitive business of the future; of the commercial find par excellence, with greater impact and economic potential, than any of the above so-called bubbles, such as subprime mortgages or pyramid scams.

Sleep, anxiety, unstoppable desire and dissatisfaction, attempt to seek solutions, which the permanent gaze to the stars has not been able to answer; these are the basic motivations of the vast segment of global customers, used by today's technomito. The triumph of technoscience indisputably lies, in having managed to seal, the gap opened by the same biological evolution: the sum of rationality to pre-existing instinct, of animal origin. The religious myth, has been replaced by the technical utopia, expressed in another way, the implant of the computer "chip" has taken over, from the ancient religious miracle. The techno-scientific promise has managed to unite two parameters, which have reached the defining range of the two historical civilizing paradigms: Eastern spirituality and Western religion.

Expressed in philosophical terms, it has sealed the union of immanence and transcendence. The first is anchored to its literal definition – remain in- and the second sign looks in another direction; the permanent desire to transfer -transcend- the physical limits imposed by nature. Both trends define Eastern and Western civilizations, respectively.

Spinoza has probably been the philosopher most concerned about this issue and its solution, though certainly well-meaning, contains an excessive dose of eclecticism; rational immanentism is still the nominalization of an unresolved basic issue, the distinction between achieving inner development and the ambition to remain. The technology referred specifically to the human being, constitutes the birth of the definitive business of this century, representing a modernized version of the ancient alchemists, with all the congenital traits corresponding to the birth of a new Religion.

Chapter VI. First Future: Technological Generalization

> "In all the great men of the science exists the breath of fantasy."
>
> Giovanni Papini -Italian writer-

Technology and its multitude of applications already make up the essential aspects of everyday life, although it seems clear that the specific relevance and weight of each of them is differential according to the importance and centrality of the different vital areas of Application. Thus, there is widespread agreement on the progress in the area of greatest vital impact, how is biomedicine and within it, bioengineering. To gain an adequate understanding of the full impact, it is necessary to divide or categorize the different groups or segments of users of biomedical applications.

wo large groups can be categorized: those we have called needy and the group of enhancers, or according to the alternative name used by a small group of specialists, oversized. The first is made up of the large number of individuals, who suffer some form of limitation, dysfunction or disease, either from congenital or acquired causes. In countries implementing the universal health format, this group is the subject of government care, both in primary and specialized care modalities. It is unnecessary to note the large volume of expenditure of this attention, in the large headings of national accounts. In those countries, such as the United States of America where hospital care takes on an exclusively private modality, the case is different.

A typical example, in which noticeable progress has been made, is formed by the sick tetetraplejics. This group mostly suffers a localized spinal cord injury, which prevents nerve transmission of brain orders to the limbs -arms and legs- causing virtually total immobility in the individual. Two types of solutions have already demonstrated a high degree of effectiveness in their application. In some cases, the incorporation of an external skeleton or intelligent exoskeleton, allows the recovery of the function of the autonomous displacement; brain orders addressed to peripheral limbs are interpreted by a computer that in turn transfers them directly to the new artificial limbs. In other cases, the solution comes from bioinformatics, by implanting two computer chips in the spine, located right at the points immediately before and after the injured section, allow the nerve transmission of brain orders, end in the desired muscles, that is, manages to rehabilitate the path of the ancient nerve connections in their entirety.

It seems obvious to point out that in both cases, a difficult period of rehabilitation is required in which the will and effort of the patient are essential, or better, definitive; in the same way, proper professional direction is required. The second user group of the technology, called enhanced and for other authors, oversized, presents a very different casuistic. This group uses technology and its advances, to improve its capabilities or potentialities, both physical and psychic of its own volition, without the obligation imposed by a dysfunction or capacity.

The use of technological applications goes through cyberimplants, mainly in legs and arms, with the obvious purpose of increasing their physical capacities. A set of individuals, forms a large set of individuals, which seeks to obtain the maximum prolongation of life. The most common method, consists of the implantation of stem cells and organs in many cases, produced from cultures of their own cell tissue, thus avoiding, the feared problems caused by rejection. The last of the notable cases concerns the use of computer devices or chips, which make it possible to connect to the internet or to some application of the area of communication.

It has been pointed out above that the indefinite extension of life constitutes a real chimera, that is, an unattainable desire. The same is not true, with the pretence about the quantitative extension of life. Simply by simple logical reasoning, it is obvious to conclude that the renewal of organs, cells and blood, will result in an increase in the duration of the life span, albeit only, up to a certain limit. It should be noted here, the existence of an intermediate point, in which individuals with dysfunction, such as acquired blindness, use necessarily expensive solutions. Similarly, applications developed for telematic and personalized medical care, through the mobile terminal, must be added to the technologically obtained media that contribute to a generalization of medical applications, although not they represent a free option, except in specific cases as will be seen below.

Thus, if there is a widespread consensus among social theorists that, the defining feature of modern Western societies, lies in economic inequality, the second part of this century will show another type of inequality, of a character much more offensive, rude, denigrate and and unacceptable: biological inequality. Expressed in another way, for a general and explicit understanding; in the year 2050, there will be two distinct groups of individuals: the citizens of first and second class. The former will achieve an estimated average lifespan, between or more than 150 years, with a high level, in terms of quality of life, while the latter will keep at best the current average age -around 80 years- suffering all the scourges and suffering imminent are typical of the inevitable aging process. Probably the most striking assertion to establish is that the cause of both categories of inequality focuses exclusively on the purchasing capacity of individuals, that is, on their economic power.

The most repeated principle, persecuted and shared by all health specialists, is stated as "the best treatment lies in adequate prevention". This laudable and accurate pretension will soon become a reality, by virtue of the use of nanorobotics. Nanotechnology, unfailingly coupled with the relentless search for new materials, with greater conductive power and application potential, will be able in a relatively short space of time, to incorporate into the bloodstream a nanofilamemt, thick thousand thick sometimes lower than a human hair, intended to detect and remove any cell, virus or microbacteria, potentially harmful to the body. This capacity will reach the cells themselves, which, as in the case of carcinogenic or degenerative diseases, begin their expansive process. It is, this time, a definitive tool aimed at achieving the desired prolongation of healthy life.

The first stage, addressed with regard to the prolongation of life, has consisted in the use of the cryogenization process, that is, freezing at very low temperatures of those who died or could do so, due to causes for which the practice scientific to use at the time, had no solution. Hope focused, on the possibility of resuscitation of bodies and minds, the moment when science could offer a solution to the original problem of disease or death.

The second and present stage consists of the beginning of the cyborg era, understood the term, as does the Royal Academy of Language (RAE) as the connection between a biological organism and a cyber-friendly. It is clear that the romantic argument present even in children's literature has spoken of such beings -captain hook or the pirate stick- Another part of scholars has used similar arguments such as that magnifying lenses are part of daily life. These arguments are not valid; in none of the cases, the add-ons are of cyber category.

The great hope derived from the experiment is that in the future, two human brains can communicate, without using spoken language. In short, the extent of the long-standing dream of telepathy seems to be closer. At a later time, the teacher's arm connected to the Internet was able to establish communication with another member of the artificial category, located in England. The geographical distance had been permanently eliminated. The range of open practical applications with this type of experimentation, result in theory, innumerable.

The current cyborg cases and the number of problems to be solved are extensive. Jens Naumann, has become the first person, to receive an artificial vision system. Dick Cheney, the well-known American politician, can live a virtually normal life, thanks to a mechanical ventricular assist device that allows you to regulate your heartbeat. Likewise, Nigel Ackland possesses the most advanced cyber arm in the world, with a digital sensitivity similar to biological. Jesse Sullivan, has been equipped with both equally artificial upper limbs, to compensate for the loss of his biological limbs; you may feel cold and warm on both prosthetics; and so the list can continue, not forgetting the most famous of them all, Neil Harbison, an artist who was born without the ability to perceive colors. An electronic eye, specially designed and fully customized, allows the conversion of chromatic impressions into musical notes, that is, you can hear the colors with the consequent improvement in your brain nerve connections.

A novel challenge faced by technological advances, it moves to the modern city and its main constituent elements, such as architectural design or transport. Technological change reaches human settlements as a generic label. Historically, the shape and distribution of cities has been one resulting from the main activity of the particular civilization to which it belonged; more specifically, its main purpose, its worldview or its great civilisational purpose. In this way, cities with an effective distribution and rationalized communication routes have been typical of the great empires, such as the Ottoman, Greek and Roman. The following example is in the walled city typical of the Middle Ages, designed for the defense of incessant warfare. At present, the dominant paradigm comes from the sustainable development movement, incorporated into the general guidelines of ecology and which translate into the "Sustainable Development Goals" (SDBs). Likewise, the growing movement of citizen organization has given way to so-called participatory urbanism.

One of the most influential trends lies in the instrumentation, previous design led by architects, urban planners and engineers, of the city self-sufficient. In addition to this reason, the second relevant label in this field is the incorporation of clean energy for the citizen supply, in all its usable modalities. One of the demands that must be faced is the result of spatial limitation, that is, the need for vertical growth, especially in the great modern cities. A first prototypical approach to this set of elements would be a skyscraper powered by renewable energies, containing parks and gardens for public use and greenhouses, intended for food production.

The limitations of usable space in large concentrations allow only the alternative of vertical growth. In this way, architects design large structures under the requirement set by smart buildings are fed by clean technologies and adding common spaces that include commercial, leisure and recreation areas next to the foo growing spaces, under the revealing vision of future self-sufficient cities.
Source: Pixabay. com

These are exactly the hallmarks of the "Singapore 2050" project and also the "Sky Build" project. In a relentless process of finding and adding sustainable elements, using both sophisticated simulation techniques and a simple "lego" - an old set of assembly by pieces-, we work on the design of efficient and combined public transport that reduce the use of the private car, to an exciting level of use, only 20% compared to current mass use. A point of special attention lies in the new concept of urban transport. In Japan, we work on single-person digitalized transport modules that cover a nearly total urban network. Different types of transport, including the train and the small car, can be combined in order to facilitate the choice of destination, practically on demand.

Likewise, the accessibility of centers of citizen interest, both commercial and cultural or leisure, must be located within a maximum radius of 5 minutes and accessible on foot. For example, the "CityScope" project created by the Media Lab of the Massachusetts Institute of Technology (MIT) uses the combined criteria of density, proximity and diversity, in order to facilitate social contact. Another example is the improvements made in the city of Copenhagen, where the criterion of greatest goodness rests on the achievement of citizen well-being.

The development of the autonomous car, is being instrumentalized together by Uber and Google, with a considerable economic investment. At this time, it does not yet show the required reliability rates, but it is already used in controlled urban circuits.

Regarding the family home, the old home automation has given way to the so-called internet of things. Through the mobile terminal, or directly by voice, household services such as appliances, general services or security, are controlled by the user.

Returning to urban planning, a unique initiative has been put up by the professional group "Urban Think Tank", in the slums of South Africa. A simple construction of two modular floors, tries to erase the difference between residential neighborhoods and the current ghettos. In this way, the efforts are made to compensate for the failure of the economic system and the management of political elites, in the necessary and progressive redistribution of space. Through this strategy, the project aims to envision the principle that "people are part of the solution and not part of the problem". Thus, the direction contrary to the previously established one can be registered, so that the design of cities directly affects, in the areas of government decision-making. The recent trend was to assume the belief that the population limit that the planet could bear was set at 1 billion people. A trend that began with the publication in 1971 of the book "The Population Bomb" by Paul Ehrlich and his wife Anna.

One more step in the goal of obtaining higher levels of citizen well-being, it is possible by incorporating sensors in various parts of the city, in order to determine the degree of pollution that exists. Thus, through the use of "big data" techniques and simple sensors, quick and effective decisions can be made, aimed at drastically reducing a danger -pollution- regarded as an alarming threat to public health.

The use of new building materials, not pollutants or energy consumers, are part of the city's consideration as a living organism; an analogy on the other hand, very common in the social sciences and technical applications in general. From this point, the initiative to use simple and biological raw materials, such as mycelium -a living organism growing in the forest- for the production of biodegradable bricks, begins.

Moreover, on a planet composed of 70% of its surface by water, it seems irretrievable that the gaze of architects and urban planners, is directed towards the ocean. Aquaculture represents a new, virtually inexhaustible form of food production. Likewise, they are in the process of design and execution, two initiatives for the construction of buildings, both on the ocean surface and a superstructure, which anchored to the sea surface and covered with several layers of glass, similar to the one used on the walls of today's aquariums, it is possible to obtain greater resistance to pressure, take the existence of a colony of 4000 people on the sea surface.

Chapter VII. Social Dystopia: The Coercive Government

> "The art of reigning is to organize idolization."
>
> George Bernard Shaw -British Playwright-

> "Despotism is impossible if the nation is enlightened."
>
> Francoise Quesnay -French philosopher-

One of the recurrently criticized aspects of this century's technoscience focuses on its recognizable commercial intention. In other words, large corporations and governments, among others, are part of the pool -segment- of potential customers of technoscientific applications. In this sense, the principle that establishes the need for a relatively low degree of opposition and social activation for a given government to exercise executive and legislative capacity emerges; the degree of opposition and social protest must be null or void. Technoscience has sufficient applications, which may be inadequately intended for the control of the population or part of it, depending on the degree of disorder or need for repression that require actions defined as potentially dangerous.

In this sense, the first group of control methods comes from *pharmacology*. The traditional medical model represents a clear example of a reductionist approach to human behavior. Linked to the prevailing typing in the seventeenth century, it contemplates bodily functioning, under the cause-effect scheme. This approach allows its maximum therapeutic objective, resides in the use of a pill for any type of disorder.

In 2009, Dutch scientists, under the direction of Dr. Merel Kindt, announced that they had discovered a "miracle" drug, called propranolol, which could alleviate the pain associated with traumatic or undesirable memories. The drug could not induce amnesia at a specific time, but it would get the erasure of memories, settling in nerve cells and blocking their normal functioning. In particular, inhibiting the presence of adrenaline; without adrenaline, the memory fades. Evidently, the smart pill has been researched for positive and medicinal purposes. Dr. André Fenton, one of the authors of a second study, stated that if this was confirmed in later work, it would be expected that in the near future, techniques based on the erasure of memories by PKMzeta would become a practice Usable.

Access to such techniques of potential control of citizens' conduct by governments has been a decisive historical practice. American economic imperialism, based on the generalization of neoliberal economic methodology, punished the South American countries, starting with Chile in the 1970s, using the University of Chicago, led by Milton Friedman; but the fact that it deserves to be highlighted, at the risk of seeming unbelievable, is that on the basis of such economic practices, they had their beginning applied in a psychology laboratory, which conducted experiments on sensory deprivation. Again, new technical applications as well as technological development show the potential duality in their use.

Using the latest advances in genetics, electromagnetism and pharmacological therapy in combination, techniques aimed at altering memories can be developed in the near future. The brain location through which long-term memory perception conversion occurs resides in the hippocampus. Dr. Theodore Berger, who leads the University of Southern California team, believes that a combination of such techniques could become a tool for the control of subjects deemed problematic.

Aggressive and asocial behavior is closely related to the human genome, particularly with a DNA sequence dubbed the "warrior gene". So the possibility of predicting this genetic predisposition can help in the prevention of hostile acts, through the development of more effective behavioral therapies. In a recent research published in the International Journal of Psychophysiology, the researchers formed a study group with 285 adolescents aged 12 to 19. The authors determined that 98% of participants who had an inactive version of the MAOA gene tended to demonstrate a high level of hostility. This gene, which is located on the X chromosome -male-, is associated with the inclination to various social behaviors, such as fighting, possession and use of weapons, and membership in gangs and gangs.

The authors of the research point out that negative emotions such as anger, aggression, hostility, anxiety or fears play an important role in human behavior. When experienced in youth, such feelings can have an important influence on overall mood, personal priorities, learning, and other cognitive processes. This behavior presupposes the presence of negative attitudes, emotions and behavioral manifestations, manifested in the form of "aggression, negativism and isolation". These results may be useful for developing individualized methods for the prevention of trends in hostile behaviour, based on the particular qualities of the person, rather than using generalized psychopedagogical principles.

Similarly, a programmed improvement of armies can be anticipated as a potential threat, taking into account the data available today. One of the institutions that has most emphonially promoted this field of research is DARPA -Defense Advanced Research Projects Agency-, the Pentagon's Agency for Advanced Defense Research Projects, which has led the development of some of the major technologies of the 20th century. With a budget of three billion dollars, DARPA has in its sights, the refinement of the brain-machine interface. Commenting on his possible applications, Michael Goldblatt, a former DARPA official, extrapolated the use of future results, in which soldiers could communicate using only thought. At that time, the replacement of damaged body parts of the soldiers would already be a common practice; results do not promise a hopeful future. Ideas that are difficult to accept, but which constitute a line of research and the daily work of the Office of Defense Sciences -a branch of DARPA-.

It should be mentioned with the intention of an objective approach, which DARPA historically bears the merit of Internet motherhood, another of the technological applications that originated in military research. The last decade of this century, it has witnessed a widespread threat, due to the use of robotics and artificial intelligence in weapons development, with the inherent concern in the scientific, ethical and political environments. The greatest concern comes from the likely use of so-called "Lethal Autonomous Robots", that is, the development of autonomous robots that could be out of human control or intervention, with the lethal ability to attack human life. Fortunately, such androids are currently used under supervision; in this way, drones and robotic soldiers do not exercise the autonomy that incorporates their denomination. In any case, in 2009 the "International Committee for the Control of Robotic Weapons" (ICRAC, in its original English acronym) was created, with the main purpose of control of this type of initiative and in parallel, to inform the international community, especially the United Nations-UN - on such activities.

The first utopia or attempt at perfect society, was starred in Marxist thinking based on the philosophical doctrine of Engels and its transformation to economic area by Karl Marx. As is well known, the fall of the Berlin Wall in 1989 ended the dream of the corporate ideal of many revolutionary groups. This fact led a former US. administration technician Francis Fukuyama to the publication of a problematic book entitled "The End of History". The author set out a conclusion not without reason. He argued that once the USSR - Union of Soviet Socialist Republics - disappeared, the Western world was left to the economy governed by neoliberal principles. Criticism of this idea has endured for two decades.

Once the main real attempt of an alternative model of society disappeared, the best utopian references come from fiction literature and also generated a wide and profound impact, both at the time of publication and at the present time , almost a century later. The utopia of fiction, that is, the vision of future society, represents in both cases a dystopia or negative and unwanted vision of the near future.

Tomas Moro in the sixteenth century, he wrote two books called utopia, which meant the origin and denomination of this literary tradition. Aldous Huxley published in 1932 the probably most famous fiction novel in history, "A Happy World". The proposal deals with a dystopia that anticipates development in reproductive technology, human cultures, hypnopedia techniques, management of emotions by means of drugs -soma-; practices that combined, radically determine the type of society prevailing. In this account, humanity is ordained in castes, where everyone knows and accepts their place in a healthy, technologically advanced and sexually free social cog. War and poverty have been eradicated and all citizens are permanently happy.

However, the paradox underlies the fact that all these achievements come at the cost of eliminating many other desirable manifestations such as family, cultural diversity, art, the advancement of science, literature, religion, philosophy and love. The title originates from a play by William Shakespeare's "The Tempest". The success achieved led the author, 35 years later, to write a review entitled "New Visit to a Happy World", specifically in 1958. George Orwell, wrote and published in 1949, the second great modern dystopia, titled "1984". In it, the constant vigilance of the "big brother", the "police of thought" and the "neo-language", determine a society manipulated by distorted information, political repression and mass surveillance.

Aldous Huxley was George Orwell's teacher, and wrote a letter to his disciple, apologizing for not being able to read the play before. He informed him that "1984" could be considered, as a continuation of "A Happy World", though he suggested, that without reaching the level of his own work. In short, he argued that the ruling class in reality resorted to medicine and hedonism to subject the population and not to a form of direct oppression, as Orwell proposed in his work.

The extrapolation to the current society of both works confers some reason on each of them. In particular, electromagnetism has been used to systematically deactivate certain areas of the brain, without the need for surgical intervention - the ancient prefrontal lobotomy. All these new tools are based on physical property, according to which, a rapidly varying electric field creates a magnetic field and vice versa; but, using the transcranial electromagnetic scanner, the activity of the areas of the brain can be disabled or reduced at will, without causing any damage. Added to this, the possibilities of dominance or control of the citizen, considered as mass, extend mainly to bioengineering, pharmacology and nanotechnology.

The figure of the "big brother" present in Orwell's work, has his royal correlation in research on the project "Brain Net" and the development of the "Human Conectoma". The connection to the Internet, through a computer chip implanted in the brain, would allow to locate and control the individuals implanted, opening the possibility of communicating complete orders and regulations, procedures and rules of action, as well as procrastinators or prohibitions, together with their possible consequences or penalties arising from their non-compliance. Another possibility in this same sense, involves the placement of sensors in different strategic points or of greater influx, in large urban enclaves.

The management of neoliberal governments, as opposed to social democratic practices, have used the projecting capacity of the evolved brain, with the aim of uniformizing a well-known form of discourse. The contrived strategies about the existence of a "built enemy" and the consequent need for the implementation of an almost permanent state of emergency, for fictional or unexplained causes, such as those attributed to cyclical economic crises - which seem to be spontaneous and unexplained generation or the inherent danger of immigration, under the pretext of the alleged threats these facts pose to freedom, work, attacks or social welfare. All of them are common causes claimed by tax governments, under the greatest of the fallacies of democracy: representative democracy. Now, moreover, of all these typical practices in the exercise of power of government, they can use all the options derived from the advances of technoscience.

With particular exceptions, the desire of any government, especially those of radical cuts, would be the possibility of developing techniques that would allow the mental control of its citizens. This attempt represents a historic constant led by the central intelligence agencies of the world's most powerful governments, the United States of America and the Soviet Union, during the ridiculous time called the Cold War.

Dr. Rodríguez Delgado, a Spanish physician, was the first to demonstrate that it was possible to control animal minds, through remote electrical stimulation, using receptor chips implanted in the brain, precisely at that time. In 1969 he wrote a book, with the provocative title "Physical Control of the Mind: Towards a Psychocivilized Society". His intention was aimed at the solution of mental illness, which at the time were treated by inhumane techniques, such as prefrontal lobotomy. However, it provoked a sense of unease, because of the clear evidence of its use with objectives extrapolated to the mind control of citizens. The press used the term "brainwashing" to refer to these techniques; expression that has endured to this day.

The Central American Intelligence Agency (CIA), convinced that the Soviets were way ahead in the investigation of this discipline -brainwashing- and also in the possible use of other unorthodox scientific methods, undertook various projects such as the MK-ULTRA. This long research began in 1953, in the attempt to explore a set of initiatives that at the time were called extravagant and heterodox. When in 1973, the Watergate scandal sparked panic in the US government, CIA director Richard Helms automatically canceled the MK-ULTRA program and was quick to order that all documents related to the project be destroyed. However, twenty thousand documents survived the purge and were declassified in 1977, under the application of the Freedom of Information Act, revealing the full scope of this massive operation.

It is now known that between 1953 and 1973, MK-ULTRA funded eighty institutions, forty-four universities and university faculties and dozens of hospitals, pharmaceutical companies and prisons, with the aim of experimenting with people who were not conscious of these practices and evidently without their consent; these practices, were distributed in one hundred and fifty different secret operations. At one point, 6% of the CIA's total budget was devoted to the MK-ULTRA project. Some of these mind control projects consisted of the development of the well-known "truth serum", with the aim of obtaining information from prisoners and suspects. Another of the techniques investigated consisted of the "deletion of memories", carried out by the US. Navy developed in a specific project, called "Subproject 54".

Evidence is handled from a large number of objectives incorporated into the project, including the use of hypnosis and a wide variety of drugs, especially LSD, for behavioral control. This type of drug has been used against foreign leaders, such as Fidel Castro. In addition, it includes the improvement of methods of interrogation of prisoners; the development of a hallucinogen with disabling effects, fast acting and without traceable trace or alteration of the character of individuals, with the aim of increasing their "docility". While some scientists doubted the validity and legitimacy of such studies, others collaborated voluntarily. Experts were recruited from a wide variety of disciplines, including visionaries, physicists and computer scientists, used in various projects, at the very least heterodox. Experiments with alternating narcotics of mental functioning; vision training for the location of Soviet submarines patrolling the deep seas, are many other examples of how citizens' taxes are squandering.

Modern biological warfare has not been developed by African or Arab states, as the American government has stated on several occasions. When the U.S. Senate received a secret report alluding to Soviet experiments with "microwave radiation," aimed at the brains of experimental subjects, rather than publishing a complaint to the international administrative media and communication, he became interested in this practice and was described as "a medium with great potential, for the development of a capable system, of disorienting or altering the patterns of behavior of military or diplomatic personnel". Fortunately, the human brain is not able to receive microwave radiation and most importantly, it also does not have the ability to decode such messages. Dr. Steve Rose, a biologist at the Open University, called this crazy project a "neuroscientific impossibility."

One of the first government attempts at the desired mental control, was the use of hypnotism. Electroencephalograms and magnetic resonance imaging show that during hypnosis, the subject perceives minimal sensory stimulation from outside, i.e. less influenced by external stimuli. Thanks to this, hypnosis allows access to some memories that are "buried", but of course, it does not have the possibility to alter the personality, nor the objectives and desires. A secret Pentagon document, dated 1956, corroborated this fact, concluding that "hypnosis cannot be trusted as a military weapon." On the other hand, administration into the bloodstream of the sodium pentotal, reduces activity in the prefrontal cortex, achieving greater relaxation and inhibition in the subjects; however, this effect does not imply an obligation to confess the truth, rather on the contrary. Similar to alcoholic ingestion, the typical "drunkvera verbble" is well known, i.e. the drugged subject acts and verbalizes in the same way that he does so under the effect of any type of drug; That is, incoherently.

Dr. Rodríguez Delgado's work was rudimentary, but they demonstrated that electrical impulses, applied in motor areas of the brain, are able to prevail over conscious thoughts, so that muscles do not obey voluntary control. In the near future, scientific and medical political institutions should control such techniques, due to their potential undesirable applications. One more example, of the continuing question focused on the debated power of the state in the face of individual freedom and above all, the use of technological advances by factual powers, whether political, economic or religious.

The ghost of a possible digital dictatorship, has been aroused in parallel with the growing trend of using the brain-chip interface. A good example is found in the "Neuralink Project" promoted by Elon Musk -tireless entrepreneur, promoter of companies such as Tesla and Space X- and cited several times. Neuralink -access or neural connection, in its strict sense- aims to undertake a vast project consisting of the massive implantation of computer chips in the brains of people, with mobility problems or paralyzed, with the aim of providing control over mobile phones or even computers. In order not to be an overly invasive technology, it has been continuously refined to achieve so-called flexible threads, small parts that could achieve great data transmission potential. The number of implants needed to achieve the goals, seems in any way excessive, since an implant of 3072 electrodes is needed in the human brain.

The possibility that such a project could become the real counterpoint of the "Big Brother" of Orwell's utopia is considered a possibility backed by history. It is obvious here that the existing possibility for governments and groups commanded by so-called "warlords", provided by different types of targeted robots, such as soldier robots or drones; in both cases, remote control by human specialists is required.

Newton's third law, called action-reaction although formulated in the context of physics, has shown extensive applicability to any type of complex system, whether physical, biological or social. Its statement reads as follows: "Any body A that puts pressure on another body B will receive a force of equal intensity of opposite direction." Coercive pressure, always excessive on citizens, either by specific governments or by a particular social system, emerges as the direct mechanism responsible for the three reaction movements, which have shaped as distinctive historical determinants of social progress. In particular, applied to the achievement of equality in human rights.

The first decisive movement, was led by the suffragettes, of exclusive female leadership, in the pursuit of the right to vote, in Britain in the early twentieth century. A few years earlier, the Bolshevik revolution had originated in Tsarist Russia under Lenin's leadership, against the dictatorship of the tsars, who kept the people unacceptable levels of poverty. The immediate result was the greatest real utopian attempt, adopting the communist model. The failure of the state dictatorship was fundamentally due, albeit, among other reasons, to the denial of the cause that had motivated the same revolution: the denial of individual freedom.

The third remarkable moment, took place in the late 1960s, specifically in 1968, named by the time it became a real worldwide social outburst. The movement of May '68, initiated by student groups against consumer society, immediately gained the support of various social actors, such as the Communist Party, trade unions and intellectuals; total people mobilized, reached 9 million. Global resonance led to rebuttals mainly on the European and American continents; Spain, Italy, Switzerland in Europe; Mexico and Uruguay in South America.

In the mid-1990s, the opposition movement erupted, which has become a temporarily stable platform. The Anti-Globalization Movement, its constituent and objective features, are set out in a later chapter.

Chapter VIII. A New Religion

> -Ethics and science need to shake hands.
>
> Richard Clarke Cabot - American physician-

> -Science for some is the great deity celestial; for others, a good cow that provides them with butter.
>
> Friedrich Schiller - German philosopher and historian-

In June 2008, the Spectrum Magazine, the only publication widely distributed by the Institute of Electrical and Electronics Engineers (IEEE) to all its members, featured a monographic issue dedicated entirely to artificial intelligence. After the initial surprise, the engineers discovered a field of scientific research that seemed to open a path to a new concept of reality, in which the possibility of the barriers between machines and humans were handled in the not-too-distant future, the barriers between machines and humans Disappeared. Machines would become, though briefly, as intelligent as human beings and they could migrate their mental contents, their consciousness and dwell inside them, perhaps in search of a probably technological immortality.

In fact, the most striking fact was that prestigious figures of scientific, technological and philosophical thought participated in this debate, some denying any possibility to the Technological Singularity and others so convinced of its proximity in the time, who said they were changing their habits of life to, by living for longer years, get to that desired time.

Although there is no specific definition of Artificial Intelligence, most authors agree that, in essence, it is to ensure that a machine has its own intelligence, a definition that supports many different conceptual approaches, because while for some authors Artificial Intelligence refers to the study of how to program computers in order to achieve objectives in tasks, in which clearly, the human individual surpasses them, for other scholars, this is a true science that attempts to create programs for machines, that imitate human behavior and compression; showing learning skills, recognition of their environment and thinking. In short, whatever the concept, Artificial Intelligence is geared towards getting machines to do jobs, where the intelligence, reasoning and knowledge of a human being are applied.

However, all these attempts were already fraudulent or real, proved fruitless, until in 1943, Culloch and Pitts proposed a model of neuron of the human and animal brain. These computerly simulated nerve system neurons provided a symbolic representation of brain activity. Nobert Wiener of the Massachusetts Institute of Technology (MIT) took these and other ideas, combining them within the same field, which was called "Cybernetics", from which it would be born, the "Artificial Intelligence".

A strong impetus in the implementation of this type of thinking, gained a major boost in 1997, when IBM's "Deep Blue" computer, defeated on its second attempt, considered by many, the best chess player in history, Gary Kasparov. Apart from other ancillary considerations, which the chess player himself stated, after the analysis of the game , the "learning" of the computer did not lead him to "make better decisions". Again, lexical contamination determines such press headlines. A computer doesn't learn, much less, it makes decisions.

First, the differentiation between the components always present in any robotic system, the hardware or physical part and substrate material and the software or computer program, which makes the entire body, usually a server, work, is forgotten. large dimensions, as in this case. The simplest example is a computer or a mobile phone, used by anyone. The components of the physical structure are determined by a set of elements that define their performance, including a processing unit, storage capacity, and operating system, as well as logically, a power source or Power. The computer program, on the other hand, contains the orders and rules of action -necessarily hierarchical- with which the machine operates. The computer, which only 20 years ago faced Kasparov, was equivalent to 4 blocks of farms and its weight is unknown. Kasparov's brain was still weighing 1450 grams.

The second distinctive entry in this case is that any next-generation computer program, at least, offers two distinct modes of refinement, for the performance of its performance; either by explicit commands in the source code, or by lines of programming code. In short, in the much-acclaimed learning process, the program undoes the alternatives that do not produce the programmed objectives and incorporates the options that do, increasing the total set of possibilities for response, in the face of a given situation . Thus, the programmed computer does not "make decisions", but applies the new existing option, with a higher likelihood of success. Nor, therefore, does it "learn", but increases formal response options.

In the rather hypothetical case, that a computer in the near future was able to access all the information on the Internet -as indicated by the hypothetical premise proper to the future dominant intelligence- the requirements necessary to achieving this achievement, simply on a physical level, would be unthinkable. The necessary physical structure would probably quadruple the size of the globe. This option seems to be, according to some thinkers, the greatest threat potential for humans. Continuing with the list of continually exposed fallacies, it is assumed in this situation of threat, a real attitude and behavior of hostility, towards human society. It's unthinkable yet ridiculous. Such a panorama would imply intentionality, motivated behavior, which clashes head-on with the basic assumptions of artificial intelligence -without resorting to the fictional laws of robotics, enunciated by the genial Asimov, in his famous fictional novel series-.

It is precisely at this point, in the decisive influence of emotions on artificial intelligence and much more specifically, the weight of chemistry or rather its lack, one of the insurmountable distances for artificial intelligence, with regard to artificial intelligence as highlighted by a renowned theoretical physicist, Michio Kaku, professor at Harvad University and scientific communicator.

As an illustrative example, a robot was recently constructed, which reproduced the nerve connections of the brain of a complex mammal, specifically a cat. Despite the level of complexity of the typical behaviors of a real cat that was able to reproduce, it lacked a mental map, which reproduced the immediate physical environment. In alternative words, the environment, the determinant of evolution, did not exist for it. More clearly expressed, he was unable to react to changing conditions and more importantly, to hunt a mouse. Human beings are defined with sufficient accuracy, as a complex neurochemical system, while an artificial intelligence body continues to mean essentially a simple electrical class system.

Also in the publications of the achievements of artificial intelligence, they influence the distortions arising from its categorization as news. This is the case, when the achievement of a robot capable of developing emotions and more specifically, of empathizing with humans was published. The central point of the success of the experience was that the robot developed positive or negative responses -approach or distance- depending on the mood of the real subject interacting with the machine. It was later discovered that the program that determined his behavior, was able to incorporate all the combinations of the 432 muscles, which form the human face and define aspects of the mood of the human interlocutor.

Clearly, robots show faster computation and storage capacity than the human brain, but under no circumstances, not currently or in the future, will they develop unrealistic behaviors, adaptive behaviors, and even less, mastery of the environment Physical. With regard to complex society, the conclusion is obvious. Successors of the ancient expert systems, artificial intelligence is unable to create abstract representations, similar to mind maps, that guide the actions of humans. A certain degree of complexity may be one of its mechanical characteristics, but it is not part of its guidelines of action.

Added to this impossibility, it is necessary to add the basic role that performs the subconscious level of brain activity in the human individual, both in situations of wakefulness and sleep. Intimately linked to the intrinsic behavior, all upper mammals regenerate their brain activity through sleep. Thus, the memories and experiences lived, are valued, cataloged and archived or enhanced. It is a grotesque image, a cybernetic body destined to vegetate, rather than to live. It doesn't seem like a desirable kind of eternity, not even for a moment.

The dream-dreaming capacity - daydreaming -, along with the longing to achieve a legitimate improved standard of living and the intuited technology capabilities rather than demonstrated, lead to the assessment of dangerous and useless possibilities. This is the case, for example, of the reduction of pain thresholds, which are found in millions of body cells. Even arguing, the potential benefits of professions acting under limiting conditions -for example, firefighters- no claim to overcome pain is justified, since the survival of any species has been based phylogenetically -traits of the species- in the perception of pain, as a signal of alarm, for survival, generating the appropriate and practical behaviors of escape and avoidance.

The real enemy, faced with the potential threat of artificial intelligence to humans, has an internal character and resides in the rules inherent in human perception itself. Anthropoformization -giving human capabilities to an object or symbol- is the responsible perceptual mechanism, in the emergency consideration of an imaginary threat, where only one set of programmed electrical circuits exist. A computer, however bulky, has no intention, nor the ability to recreate a concept as simple for any animal as the immediate environment; it is not able to symbolize or abstract; can't even interpret an algorithm that describes the simplest emotion. In short and expressed with the utmost clarity: it is a machine, no need for data it can store; a storage and calculation mechanism, but only that. Its basic usefulness is to serve as an auxiliary instrument, for a wide variety of human activities. It could accumulate all the significant data of human history, without deriving some kind of typical and common emotions that have defined this historical process, such as selfishness, pride, greed or contempt.

An advanced cybernetic, could visualize millions of images, showing the most varied emotions and would be unable to make a judgment, about a process as simple as the intensity or depth of that emotion; a response that a pet emits on an instinctive basis. This is and will be the case, even if it takes a thousand years and computer programmers continue to write millions of lines of code. You can never see a car crying because it doesn't have gas; may ignite a pilot and even vocalize the fact, but it will be humans, the subjects who will evaluate the consequent options; they will decide the course of the subsequent action, fill the tank immediately if they are traveling, or do so the next day if they are in a hurry to get home and rest. Likewise, no artificial intelligence, will feel the need for company nor will it be able to tell the sensations derived from a work or sports achievement, let alone taste a reserve wine. The robot with greater intelligence would not have been able to survive a single day during a clash of medieval wars, although it may be programmed to cross a traffic light the moment the light emits the green hue.

In this case, the determining lack of artificial intelligence -sometimes contrived- has to deal directly with the parts of the brain, typical of animality, such as the reptilian brain and the brain called mammal; the two regions, which predate the appearance of the human brain, located in the prefrontal lobe, responsible for association functions and complex thinking. Mammals superior to the human being, distinguish in a primarily intuitive way, that is, objectively, the awareness of their individuality with respect to others and the environment. In the human case, consciousness is the result of a continuity of memories that make up its present individuality.

Considering how the process common to higher animals and the human being, conforms to individual consciousness, can be defined as the temporal succession of a set of emotions, which involve associated image or memory of situations determined; in this way, the picture is definitively clarified. That is, the Orthodox vision of formation of consciousness, has been defined as the process of categorizing memories, concreted in images, which necessarily involve an emotion, more or less intense, as an indispensable condition for their inclusion in the functional categories of remembrance. The archiving process is carried out according to the centrality or importance it possesses, a certain experience for the receiving individual. If factor inversion is adopted, the cognitive process responsible for memory consolidation, basis of consciousness, the resulting format, comes closer to the actual process. That is, the formation of consciousness, as a succession of emotions, which involve an associated image or memory. This is the case in higher mammals.

This approach would explain the fact that it is easy to find and pervasive that in the long conversations that usually take place in social gatherings, personal positions on core or important beliefs, such as religion, politics or marriage, after hours of listening and producing more or less practical or logical arguments, remain the same at the time of returning to the individual home. Attempts are made to modify through reasoning, deep and ingrained emotions. This is only possible, under certain conditions, those that correspond to artificially produced situations and which do not depend on the experience vital to their training, for example, a situation of exceptionality defined by a particular government. That is, a reality constructed artificially and not sensorily experienced.

In this way, the greatest promise generated around artificial intelligence is reached: the indefinite duration of existence, by doubling a human brain with its own consciousness into an artificial support, mainly robotic. Although technically possible, the very evolutionary nature and its essential features, including prominently, the emotional substrate of consciousness would make this option impossible, assimilating the perception of the integrated individual self to a succession of images, similar to the viewing of a silent film; without future emotions or experiences. In addition to this, the predictable isolation of this hypothetical form of existence would lead to progressive brain deterioration. The stimulation of the environment allows the activation of the cerebral cortex and in this way, the motivated functioning of the brain and as a consequence, of the body in general. Expressed in common language: the cybernetic individual would go crazy.

Technoscience has led to a new form of religion, as it shares with it a number of common elements. It has managed to divide more than unite the Western social group, as was the case with monotheistic religions. It has aroused in the population an identical reaction to religious fervour and in the same way as religion, they have defined an area of popular worship that, because of its profound significance, is intended to be financially monetized. The similarities between the two beliefs are likely to be greater than their differences, if the methodology used to achieve the promised ends is overlooked. It is certainly curious that after crossing so many criticisms, between religious followers and the agnostic defenders of scientific practice, the latter elevate to the category of maximum promise, a physical and biological impossibility: life Eternal. Even though there are essential differences in how it is obtained. In the case of religion, it requires metaphysical instances and extra-corporeal and magical methods.

Science offers proven methodologies, both conceptually and operationally, but in essence, the ultimate goal is identical. Technoscience has introduced, among its substantial promises, the aspiration to an indispensable future substitute for religious predestination, fully and almost immediately satisfying the ancient human need for anticipation or prediction of the future, as they did at the time, ancient astrological and religious beliefs. Artificial intelligence has managed to generate a mystical halo around it, occasionally presenting a new elaboration of the traditional, divine and demonic deities; the use of technological applications leads with absolute certainty, to the heavens of the prolongation of healthy living; their undervaluation or contempt, entails suffering, pain, old age and unnecessary death for the individual. The technological myth has established new stages of the journey to eternal happiness and has also built new temples of worship. These are the departments of large universities and private research centers. The climate that is breathed in the university libraries, emanates a fervor and respect for quasi mystical.

Indeed, in addition to research centers, who are fully specialized in mono-thematic epigraphs such as Google's Calicó, which studies anti-aging techniques, large universities and traditional institutes such as Harvard University, the Massachusetts Institute of Technology (MIT) or the Max Planck Institute in Munich, a pioneer in biomedical research, continue to be at the forefront of basic and applied research. In Spain, the Center for Biological Research (CIB) belonging to the National Research Council (CSIC) includes among its main research, the disciplines of structural and cellular biology, biomedicine, agriculture, biotechnology, environmental sciences and, of course, chemistry. The research programs of the CIB - Center for Biological Research-, are organized in five departments - Cellular and Molecular Medicine, Environmental Biology, Chemical Biology-Physical, Molecular Microbiology and Biology of Infections - and a new IPSBB interdepartmental unit -Integrated Protein Science for Biotechnology and Biomedicine-.

Associations between disciplinary centers represent a common feature in global research. So this summer, three European centers, leading in research in plant biology and agrigenomics, have signed an agreement that will strengthen their relationship at the scientific and institutional level, promoting the vision of the European Research Area. In this way, an alliance has been born between the Agrigenomics Research Center (CRAG- located in Bellaterra), Barcelona, Spain-, the John Innes Center in Norwich (United Kingdom) and the Max Planck Institute for Research in Plant Improvement (MPIPZ) in Cologne -Germany-.

In the United States, Boston has become a second Silicon Valley. At the heart of MIT is the Picower Institute for Learning and Memory, which according to its corporate slogan, is changing people's lives as it is now known. The Brad Institute is considered the best genetics institute in the world. The Koch Institute is among the best in cancer research; The Ray & Maria State Center lab qualifies as the leader of artificial intelligence and finally, in the Media Lab, the great human experiments are conducted in their relationship with the machines.

This geographical area, united by the urban metro, concentrates the largest concentration of Nobel prizes in the world; at MIT they have won 78 awards, Harvard University, 44 and Boston University, 4. Thus, a student at any of these centers has a high probability, that one of his teachers is a Nobel laureate and more importantly, it is very likely that the healing of cancer or Alzheimer's will be achieved in one of them.

Iñaki Gabilondo, well-known journalist and analyst from our country, presented a television program of scientific dissemination, oriented to the future development of science, entitled "When I am gone". In one of the programs, he recounted his experiences of a recent visit to Silicon Valley. He put out that the deeper feeling that had inspired him the tour of the various companies located in the place, had been the "deep sense of confidence in the future that emanated in all the centers visited". Normal, dear Iñaki; the future is them.

John R. Searle proposed the concepts of weak and strong Artificial Intelligence. Weak Artificial Intelligence is defined as the rational modality, focused on a specific and determined task, covering all automations that execute operations in a specific field. Its field of action is limited to a narrow margin such as industrial robots that manufacture all or part of consumer goods, industrial goods, automobiles and other products. Therefore, it is often referred to as "Narrow Artificial Intelligence", as machines could act as if they were intelligent and pretending to be smart, they would choose to outperform their programmers, in restricted procedures due to their high speed computing and memory capacity; but always in concrete spaces of action and without having any capacity of the general knowledge of its environment, but only, the functionality necessary for the execution of the tasks entrusted.

Instead, the defenders of "Strong Artificial Intelligence" claim that all the components that characterize the human mind, have their counterpart at the time of incorporating a computer program into the artificial circuits. This assertion could lead to conviction that machines are capable of producing thoughts, just as a human. It would suffice to do so by implementing the right kind of programs to sequence thought and acquire awareness of their actions. To do this, autonomous routines and mechanisms would be programmed, in the reflective, analytical, control and self-critical processes, leading to the correction and optimization of their task. To achieve these goals, cybernetic saiveness should possess a human-like worldview; in short, to achieve the commonly denominated, judgment or common sense.

It is clear that philosophical and ethical issues arising from artificial potentialities focus on the area represented by "strong" artificial intelligence. This includes the illusory speculations already discussed, about the possibility of creating artificial "minds" that can evolve on their own, until they reach what has come to qualify as super-intelligence. The projection of this scenario typical of the fiction of the last century, would provoke outcomes so happy, as a humanity practically liberated from work, necessary until that moment, for its survival. Similarly, a super-intelligent entity could, by itself or guided by third parties, exercise total and absolute dominance over humanity, becoming the ultimate cause of the destruction of human civilization.

These scenarios, like the many other recounted, of similar scientific tasting and potential for extrapolation, simply fall within the realm of factual impossibility. The arguments have been set out before. The responses to these dystopian visions, issued by scientists belonging to a wide variety of specialties, although united by logic and broad scientific knowledge, have been multiple and varied. Specific arguments, contrary to the fictional potentials of the artificial intelligence phenomenon, vary in their degree of application and in the generality of their construction. There are a wide range of objections to the generalization and applicability of artificial intelligence. The first one takes on a mathematical or logical character. Basically, it states that if a statement expressed in any language or code is correct, it is necessarily incomplete. It does not appear that in this case, it demonstrates its capacity as an argument for opposition. This theorem refers primarily to scientific or logical statements and is called the "principle of incompleteness" formulated by Godel, whose reading we do not recommend, because of the danger that it could exert in the loss of the mental balance of the reader.

The argument of consciousness is equally relevant and expresses the impossibility of the machine for reflection or feeling, regardless of the complexity of the task accomplished. In short, the machine is not able to experience joy for its achievements or sadness for its failures. The so-called argument of the various disabilities, has been developed in different ways, but in general refers to the inability of the machine, in the attempt to reproduce essentially emotional human capacities, such as savoring an ice cream or experiencing the feeling of infatuation.

The objection raised by Lady Lovelace -a famous British mathematician and writer who died in 1852- is obvious; a machine has no creation capacity, its capabilities are limited to order fulfillment. More interesting from a scientific point of view, it shows the supposed argument based on the continuity of the nervous system. It is configured as a continuous fabric, for this reason, it can never be simulated by discrete devices - computers process information in discontinuous categories, specifically in 1 and 0 - such as digital computers. Greater sophistication is found in the reasoning of the informality of behavior, condensed into the impossibility of developing a set of rules, which allow the description of all the behaviors performed by an individual, in the whole of possible circumstances; similar to the principle of incompleteness set out by Godel.

The hypothetical human brain-cyber livelihood cog has originated the hypothesis called "God Equivalence Hypothesis". In its formulation of maximum idealization, this hybrid being would consist of a complex system of recursive life perpetuator character. In reality, it represents a new type of computer simulation in which the machine acquires the capacity to fully develop its knowledge potential, to the point of acquiring the power to control space and time. This virtually infinite potentiality would allow the machine, among many other options, to introduce changes in history, placing itself at the origin of the times, that is, at the origin of the Universe. Again, several paradoxes combined in the repetitive irrational hyperbole; time travel coupled with the power of creation of divinity. This is precisely the end of one of Isaac Asimov's most famous novels and not by chance. This type of conclusive reasoning still belongs exclusively to the realm of science fiction.

Technoscience in general, as currently formulated, can be classified as a phenomenon of political genetic consequence, since it involves the automatic division of opinion on this subject of the social whole, on separate and general sides, Opposite. It has an impact on social analysis that can be described, as do many other doctrines in other fields of knowledge and social theory, from a binocular approach. One of the significant drifts, framed in the vast panorama of technological innovation, is represented by the "Transhumanist Movement", whose principles form a true belief of clearly escaphological cutting, understood this in strict sense, that is, as a conceptualization of human nature and its place in the universe. Stated on its website - still under construction - this large collective defends the right to use improvements of any kind, existing in the current scientific landscape. A test of the doctrinal mantle with which the transhumanist proposal is covered is found in the publication of the work "The Age of Spiritual Machines" by Raymond Kurtzweil in 1999, one of the well-known leaders of the new alternative technoscientist.

The theoretical assumptions included in its program of action, present a multiple form of hybridization, between human and machine. It forms a new doctrine, understood as a complete system of beliefs, whose main objective lies in the elimination of the disease, the delay of aging and ultimately the achievement of immortality. Aubrey de Grey, a well-known British gerontologist, has sufficiently described the aging process, providing solutions for its elimination, based on cell renewal techniques, including the grafting of stem cells or stem cells.

Initiatives concerning the attainment of physical immortality, use without exception, some modality within the general concept of mental transfer, known in English as "uploading"-, in direct reference to the process of "uploading" or "burdening" possible of the human mind, in this case, in the form of data to virtual platforms located on the Internet. However, some well-known theorists such as Bostrom and Sandberg, have proposed an equally unworkable alternative. The ultimate goal focuses on a total hybridization of the brain and a cybernetic agency, usually a robotic application. The way to achieve this end is based on the first possible format, on the scanning of the molecular and atomic structure of the human brain. In a second moment, a simulation or virtual emulation of the brain would be designed, which would be moved to an artificial structure, thus ending, with the threat of human body fragility. One more link in this line of reasoning, it underpins the main proposal of the so-called "Global Future 2045 Initiative", founded by the Russian billionaire Dimitry Itskov. All the options proposed, go through different degrees of integration between brain and machine. It is obvious to note that the current state of technology, although advanced, continues to clash with the constant barriers derived from human dreams.

If this future picture is achieved, the philosopher David Chalmers defines only four possible scenarios for the transhuman world: extinction -he human being as we know it would cease to exist; isolation -segregation between conventional humans, bio-technologically altered humans and artificial superintelligences; inferiority – humanity as an inferior species susceptible to enslavement or elimination and integration- fusion of humanity-technology. It would be in summary the beginning of the so often proclaimed time of technological singularity.

The globality of the impact of technoscience reveals that it has become the new universal religion. The religion of the 21st century rests on cultural materialism; the totem object of worship is money, unfailingly linked to the pursuit of power. Thus, those groups that cannot access the required domain level lose their longed-for reference. The real power elites move through friendism and clientelism; they constitute the place where everything is easiest, the authentic "crystal sphere" in Sloterdinjk terminology, from which the world is observed, without being affected by the changes and jolts that occur within this privileged circle. The large middle social class collective only has the academic training to attempt the assault on the great and illusory promise of the current system: social escalation. However, it is unaware of the anticipated end of that it will never reach the true summit. At one point, there is an authentic glass ceiling. It belongs to the elite by birthright with a fortune of years and tradition. The new rich, have their own distinct spaces and dimensioned by traditional elites.

The ritual adopted by technoscience faithfully follows the path marked by the worldview encumated to economic neoliberalism, canonized as a universal practice of the Western economy and publicly advocated by Ronald Reagan in the US and Margareth Thatcher in Britain, under the joint but false motto of "are not another way" - there is no other way. Basically, it consists in the implementation of global businesses under the triumph of economic neoliberalism and its principles. The relationship between power and money is not automatic but consistent, since it entails a social escalation with new social relations in turn, with a higher level of power; And this is and has historically been conceived as the domain over others, often understood and configured as mass.

If the promise of eternal life were to take the form of a classic film, today's religion would represent a two-dimensional vision, while technoscience could be viewed in a three-dimensional -3D format, with a greater appearance of reality. But both do not cease to have an ideal meaning; they set up a virtual story, at this time unreachable. Both eidetic systems -religion and technoscience- show a clear conceptual mimicry; same promise, identical goals. Technoscience represents an updated isomorphism of religion; with different methods, a digital clone of religious practice; in short, the emergence of technomyth.

Historically, the influence of the perceptual parameters of humanity, susceptible to the influence of myths, both ancient and modern, has been found. In this sense, the new economy has set its great strategic objective in the conquest of the mind. Manuel Castells, is a well-known and prestigious sociologist, economist and writer, author of the "Society Network"; Harvard University professor for a long time, shares a department with two prestigious American cognitive psychologists. In an interview for a Spanish private television network, he stated that his vision of the economy and markets, had evolved progressively, to the point of conclusion, that in virtually all the relevant fields of public activity, both economics as a politician, the ultimate goal, focused on the conquest of the human brain.

True reflection of this appreciation, a remarkable change has occurred with some resoundingness, in the top positions of the world exchanges, index by the world economy. At the beginning of this century, the first positions in the stock market ranking were occupied by service or production companies. Only one technology company ranked third: Microsoft. Next to it, an energy company, Exon and a large banking corporation, City Bank. For relatively few years, the top places the classification, on the other hand absolutely forced and contrived, have been taken practically by assault, by the four companies that group the acronym GAFA, i.e. Google, Amazon, Facebook and Apple, while Microsoft holds third place. The term is commonly used in media, in meetings of broadcasters, in online forums or in legal texts, although not in the common language.

Amazon, Facebook, and Google don't need physical distribution channels, on the other hand, if they're needed by Apple, because it's the only one that markets a physical product. The new economy does not offer services only; its true business, remains information and its handling, giving full meaning to the old principle that reads "information is power". The metadata provided by the cooperation of the consumer through the acceptance of "cookies" - user data of electronic devices- allows the new international capitalism to be adopted, a friendly and democratic face. Nothing further from reality. The European Union -EU- has fined both technology companies -Facebook and Apple- under the charge of monopoly practices, without daring to do so for unethical or improper conduct. The four companies take the top spots in the world and their economic value at the moment is incalculable. The real technology really remains financial engineering, just like at the end of the last century. Tax practices remain socially irresponsible and legally unlawful.

EU legislation is very strict on monopoly, competitive practices and taxes and its obsession is on the attempt to standardize all 28 countries that form the union, all with cultural, language and social differences. The GAFA group adopts global decisions, with identical effects in Europe, Asia or America. The European Commission sanctioned Alphabet, the Google search engine's parent company with a fine of 2,424 million euros. The legal justification led to an abuse of a dominant position in the product comparison service. It is the largest European fine in history to a single company. Google was specifically accused of leveraging its overwhelming dominance position in internet searches, to artificially promote Google Shopping appearances, as well as eliminating potential competitors.

The dominant societies of the capitalist system at the end of the twentieth century, restructured a new technological revolution of historical proportions, the core of which refers to the technologies of information processing and communication; the formation of an international economic system that functions as a real-time unit, supported by a technological infrastructure that enables such concurrency; and a process of profound socioeconomic restructuring at the global level, which has laid the new foundations for capital accumulation, as well as for political legitimacy within each nation, while imposing significant social costs on all Countries.

Thanks to the new infrastructure, provided by information technologies and accumulation, the global economy has been able to make the conversion truly global. This new modality directly affects each and every one of the processes and conforming elements of the economic system. Despite the persistence of protectionism implemented by various governments around the globe, with the aim of protecting national economies and restrictions on free trade, markets in goods and services continue their expansionary process, the light of the old liberal principle: "Laissez faire, laissez paire" - let go, let go- The information economy may differ in its dynamics of action and in its contents of the productive economy, but it is not contrary to its functional logic, rather on the contrary. The industrial economy had to become informational and global or would have collapsed. In this new capitalist context, the informational term indicates the attribute of a specific form of social organization, in which the generation, processing and transformation of information become the fundamental sources of the productivity and power.

The so-called third revolution is increasing the globalization of markets, the internationalization of production and global competition. Globalization and ongoing technological change are restructuring the international economic order, through an exaggerated and fragile dynamism of global financial markets, through the increasing expansion and diversification of investment foreign direct and new exports of services.

These conditions call for new ties of cooperation between the State and the productive sectors, leading to greater investment in the field of innovation and research; in the development of greater endogenous technological capabilities and the elimination of external dependence on the various world governments. States that do not attend to this new reality, investing decisively in the widespread formation and training of the new society, will face a path of marginality and secular delay, probably of irreversible sign. The precise strategic changes in today's society focus on some new developments, which can be developed in light of new technologies, such as investment in clean and renewable energies or digital transition; but the investment that has persisted unchanged since the beginning of large societies obviously lies in the education of citizens. Even Tony Blair -a former British prime minister -unsuspecting of defending social democratic positions, publicly touted at the beginning of the century the ultimate goal of the society of the future: "education, education, education."

Hours after the Cambridge Analytical scandal, Facebook suffered a $37 billion stock market crash and its owner, Mark Zuckerberg, was forced to testify before an ethical commission of the US government. He publicly acknowledged that he had allowed the collection of private information from millions of users, through the use of an alleged personal habit survey. The results were published by the leading American newspapers, the New York Times and The Observer and used for the segmentation of election campaigns in the last American presidential election. In 2016, Donald Trump was elected president of the United States.

Critics, mainly from the theoretical economy, who deny that both modes of economic action -financial and informational -constitute new forms of capitalism, should analyze these facts in detail and if they consider it should, modify its conclusions.

Chapter IX. Children of a Lesser God

> "Science and technology revolutionize our lives, but memory, tradition and the myth frame our response."
>
> Arthur M. Schlesinger - American historian and critic-

> "The most beautiful experience we can have it's mysterious. It is the fundamental emotion that is located in the cradle of the real art and true science."
>
> Albert Einstein -Austrian physicist-

It would be right to say, that the prevailing feeling in future parents, at the close moment of the birth of their first child, is to conceive of the desire for the new family member, of all that he can obtain and that contributes to his future and also happiness and blissful existence. The first impulse is the desire for children to possess naturally all the attributes necessary to obtain these achievements, but for some time, the possibility of achieving them through scientific practice is considered a real option. Parents would like their descendants to look like a demigod in practice.

In the case of Christian doctrine, Jesus of Nazareth inherited as much gift as possible; divinity with all its inherent attributes. The same was not the case with Alcides, son of Zeus and known to future generations as Hercules, the demigod. According to the legend Zeus, the highest God of Greek mythology, was punished by his wife Athenea to have a child with a human woman, Alcmena, because of his constant infidelities. According to legend, Alcides possessed a miter, but apparently his intellectual qualities did not respond to the same magnificence.

A few years ago a regular frequency survey was conducted on the degree of happiness of citizens. The data analysis officer was very surprised, to find that under no circumstances, one of the interviewees, would have chosen the option to increase their degree of intelligence. In 2013 Nick Bostrom and Carl Shulman, two researchers from the Institute of the Future of Humanity at the University of Oxford, were commissioned by the journal "Global Policy" to investigate the social impact that the empowering of intelligence could awaken in the citizen collective.

In principle, they concentrated on the selection of embryos for in vitro fertilization (IVF), since through this technique parents can choose the embryo, which they wish to implant in the conception of their future child. According to his calculations, the choice of a supposedly smarter embryo, among the 10 available, would increase the child's IQ by about 11.5 percentage points, relative to the one that would correspond to him, leaving the process to chance. Following this reasoning, after 10 generations, the descendant of this genetic line could reach a ratio of 115 points, above the original ancestor, that is, would have the IC similar to that currently considered, corresponding to a Genius. Even if it were an optimistic view, the article concluded, that using embryonic stem cells that could become an egg or sperm in just six months, the results would be faster.

It is really an optimistic view, since to make the aforementioned genetic selection, it is necessary to know what is the genetic basis of the intelligence and at the moment, it is unknown. There are several reasons, determinants of this ignorance. First, the different cognitive modalities -spatial, complex reasoning, spatial, analytical and others- are determined multi-genetically, that is, they owe their function to more than one isolated gene. In addition to this, interaction with the environment is crucial. Environmental factors, as geneticists often seem to forget, take a decisive role in the final outcome of the intelligible degree of any individual.

In this regard, Stephen Hsu, deputy director of research at Michigan State University and co-founder of the BGI Cognitive Genomics Laboratory -the former Beijing Genomics Institute- in a 2014 paper noted that there are possibly about 10,000 genetic variants that influence the final determination of intelligence. Despite the estimated considerable number of incident genes, some researchers say that such a number can be handled in their entirety within 10 years. At the same time, they are sure they can get results by choosing those genes already known. Something similar happened in the formulation of the smallpox vaccine. At the moment, experimentation using the CRISP technique is reduced to its application in animals, due to ethical and of course legal restrictions, for experimentation in humans.

As for the possibility of increasing physical capacities, as described above, it inevitably requires cybernetic implants, mainly in arms and feet, forming a hybrid being -cyborg- or a bionic man, in greater or lesser Degree. But the traits usually admired are found in fictional characters, protagonists of films and ancient comics, that is, in superheroes. These fictional characters often have the ability to master the physical objects around them or to read the minds of others, that is, exercise telekinesis or telepathy.

Telepathy is currently the subject of intense research at universities around the world, where some teams of researchers have already managed to use state-of-the-art sensors to read individual words, images and thoughts, produced in the Brain. This fact could substantially alter the mode of communication with victims of strokes and accidents, who find themselves "trapped" in their bodies, unable to verbally articulate their feelings, except by winks or minimal gestures; unfortunately, this kind of research is fruitless. Telepathy could also radically change the way it interacts with computers and the outside world. Scientist´s at IBM claimed the possibility of mental communication with computers, foregoing, the use of mouse and voice commands.

An optimistic vision of the future - again - contemplates these possibilities as the option of using mental power for telephone calls, credit card payments, motor driving or the creation of beautiful musical symphonies. The apparent possibilities are varied and apparently many actors in this field, from the computer giants to the teachers, to the companies that develop video games, the music studios and even the Pentagon, direct their interests, converging on the use of this type of technology. True telepathy, which appears in fantasy and science fiction novels, is not possible without outside help. As is known, brain capacity has a main characteristic derived from its electrical functioning. In general, when an electron is accelerated it emits electromagnetic radiation; identical event, occurs with electronic particles that oscillate inside the brain; produce radio waves. But the signals are too weak, so that other minds can detect them, and even if they could be achieved, understanding their meaning would be a task with a high difficulty index. Evolution has not provided humans with the necessary capacity for decryption of radio signals, but computers are able to do so, just as dolphins interpret a complex language based on the use of wavelength frequencies.

Scientists have managed to obtain approximate readings of a person's thoughts by using electroencephalographic scanners -something similar to a brain x-ray-. Since neurotransmitters such as dopamine, serotonin and norepinephrine, make it easier to control the flow of information that runs through the multiple grooves of the brain structure, they have a powerful effect on thoughts and also on moods. The scanning technique allows an interesting approach to the complex cognitive dynamics, estimating the interaction of the different interacting forces.

Telekinesis as a real driving option offers even less hope. In the future, even with an external energy source that amplifiers the potential of thought, it is unlikely that people with estimated or assumed telekinetic powers could achieve the minimum displacement of objects as simple as a pencil or a cup of coffee. As you will see in a next chapter, there are only four known forces, which govern the universe and none of them is able to force the movement not determined by natural laws of objects, regardless of their size, without the contribution of a source of external energy. Magnetism might approach this attempt, but its influence is limited to objects with magnetic properties; objects made up of plastic, water or wood can pass through a magnetic field. The mere levitation, a trick that is part of any circus spectacle, escapes the technical capacity. In conclusion, even with the use of an external energy source, it is practically impossible for an individual, supposedly gifted with a predisposition for telekinesis, to achieve movement at will of the objects in his environment. However, a type of technology likely to make progress in this field has been reported. This is the ability to transform one object into another.

This technology is known as "programmable matter" and is the subject of intense study at Intel Corporation -the company responsible for manufacturing most processors inside our computers. The idea behind programmable subject matter focuses on the creation of objects composed of small self-programmable spheres –called catoms– which are actually microscopic-sized computer chips. Each of these units is controlled by a wireless connection and can be programmed to vary the existing electrical charge on its surface and link to other atoms in different ways. If electrical charges are programmed in a certain way, the crates are joined to form another larger object, such as a mobile phone.

At a later time, by pressing a simple button, the programming of the catome varies and these are reordered composing a different structure like a laptop. In short, it is the possibility of programming the hardware -physical part of a system- in the same way that the software -the information component of the same- is currently programmed.

This information allows the physical instrumentation of a direct interface -a simple face-to-face- between the brain and the computer; with the capacity to investigate the possible control of any object in the physical environment. This fast-growing field of research is known as the Brain-Machine Interface (BMI) and key technology focuses on the requirements of the computer, which should be able to get recorded memories, mind reading, video recording of dreams and also, the desired telekinesis; it is the analysis of situations that have occurred in the past, in order to produce completely new situations. This principle is identical in its approach, to that used in advanced artificial intelligence learning programs.

In a recent but well-known experiment, scientists at Brown University placed a small chip -dubbed Braingate- wired to a computer on top of the brain. The patient's brain signals were transmitted through the machine to an equally robotic mechanical arm. Simply by thinking, the patient has managed to learn slowly and progressively, to control the movement of the arm, achieving, for example, to grab a bottled drink and bring it closer to her mouth. For the first time in her life, this woman has experienced the feeling of having some control over the world around her. The paralysis also made speech impossible, so that he had to communicate the emotion experienced, through the movements of the eyepieces. An electronic device has also been designed to track the movement of your eyes and their translation into written messages. The answer to the question about his emotional state, after years reduced to his body shell, was literally unequivocal: "Euphoric!"

Professor John Donoghue and his colleagues at the Universities of Brown and Utah have manufactured a small sensor that acts as a bridge with the outside world, to be used with those individuals, unable to communicate with the immediate environment. The brain-gate has opened a new door to a wide variety of state-of-the-art neuroprosthethics, already enabling, that a person with some kind of paralysis can move artificial limbs through the mind, in addition to communicating directly with their loved ones and a small number of other individuals. The first version of this type of chip tested in 2004 was designed so that patients suffering from paralysis could communicate with a laptop. Soon after, these same patients surfed the internet, read and wrote emails, and controlled their wheelchairs. More recently, a neuroprosthethic was incorporated into the glasses of the world-renowned cosmologist Stephen Hawking.

As an electroencephalogram sensor, this mini-device connects your thoughts to a computer, allowing you to maintain a certain type of still-restricted contact with the outside world.

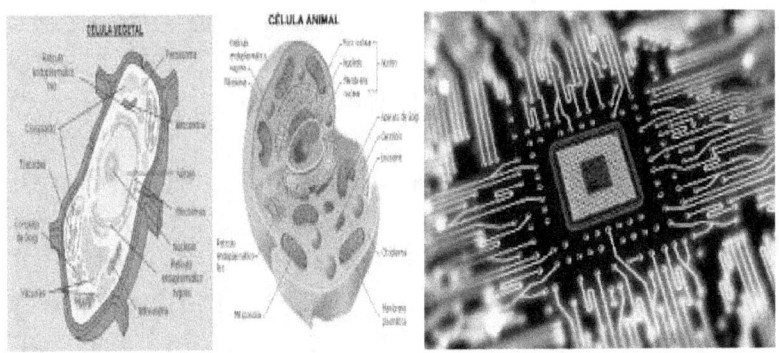

The human cell interface, in this case mainly of brain neurons, represents a qualitative leap from the second phase existing in human or cultural evolution of an exponential nature; aims broadly to improve human capacities, as well to eradicate the disease. Logically, its final use will determine the objective benefit or possible harm to future society and even to inmediate society. Source: Pixabay.com

The brain-machine interface has already penetrated almost completely into the youth market, in the form of video games and toys that use electroencephalography sensors, allowing mental control of objects, both in the virtual plane and in the real world. The three-dimensional glasses used in conjunction with stereo helmets, are the accessories used in the current game applications, offered by virtual reality. The development of a brain-net -a network of brains- will also boost the multibillion-dollar global entertainment industry.

In the 1920s, the technology of sound and light tape recording was perfected. This initiative triggered a transformation in the entertainment industry, which went from silent to sound. For much of the twentieth century, the combination of image and sound remained unchanged. But, in the future, the entertainment industry could undergo a new transition and enable the recording of perceptions, coming from all five senses, including smell, taste and touch, as well as a whole range of emotions. Telepathic probes would be able to handle the entire range of senses and emotions circulating through the brain, generating a complete immersion of the audience in any type of scheduled history. While viewing a particular genre of film, such as romantic or action thriller, the viewer might experience identical sensations as actors. It might be possible to perceive the scent of gunpowder, the panic of the victims in a horror film or the taste of the hero's victory over the villain.

Virtual reality will undoubtedly become one of the most profitable future business. Its possibilities as a practice of fun, evasion and probably therapeutic are immense. Source: Pixabay.com

It has only been fifteen years, since the emergence of the MRI tool and other sophisticated types of brain scans, to achieve the active connection of the brain with the outside world, that is, the realization of brain-world interface. This progress, with increasing acceleration, has been possible by virtue of the widespread understanding by physicists, of the mechanisms of electromagnetism that govern the electrical signals that travel through brain neurons. The mathematical equations of James Clerk Maxwell, have been used for the calculation of the operation of antennas, radars, radio receivers and microwave towers and are the cornerstone of the basic technology of resonance imaging tools Magnetic. It has taken centuries to gain the definitive understanding of electromagnetism. Today, three-dimensional, full-color images of human thoughts and emotions can be observed.

However, the answer to the advancement of this technical specialty can be governed by Moore's law, the basic principle of which states that global computing power doubles every eighteen months. In this sense, it is surprising that today's smartphones have a higher computing power, which NASA could use with their computers, at the time it brought three men to the Moon in 1969. Computers now have sufficient capacity to record the electrical signals emanating from the brain and partially decode them, in an understandable digital language. This allows to establish a direct interface between the brain and computers, thus controlling any object of the immediate physical environment.

Chapter X. A Complex Demon

> "Science commits suicide when it adopts a creed."
> Thomas Henry Huxley -British philosopher and biologist -

It has been exposed in the first chapter, the use of models and theories to explain the observed reality, starting with the two most notorious and relevant manifestations, namely the movement of objects and the functioning of the universe. Reality as a complex phenomenon cannot be fully explained, due to the cognitive limitations of the human being. For this reason, both theories and models, mean reductionist abstractions of the world. The movement on its largest scale, corresponding to planets and galaxies, was explained in the 18th century, using Newton's gravity equation. The perception of the universe as matter extended to natural systems and also to philosophical and social proposals. A greater degree of accuracy and complexity, was introduced by Einstein collecting in what is probably the most famous and decisive mathematical equation in history, the equivalence between matter and energy and indirectly, the presence of a constant, the square of the maximum possible speed achievable in the universe: the speed of light.

The implications of this formulation continue to dominate the perception of reality at any of its levels of study, observation and research. The main nuance is that obtaining matter from energy is virtually impossible. The demands of the amount of energy needed for obtaining a small amount of matter are immense. For this reason, the main generators of large volumes of materials, in practice are reduced to the cores of the stars, subjected to millions of degrees of temperature. From that moment, the components of the materiality are interpreted and sometimes reduced, to a compendium or unit of energy.

A step mostly encompassed than that offered by mathematical formulation, it was proposed with the "system theory" that contemplated the different levels of objectivity, as a set of elements that established an interaction between them. That is, it implies that the whole is more than the simple sum of its parts. This new optics, allows the observation and more complete understanding of physical phenomena. For example, the contemplation of the Milky Way, makes it possible to see the role of the Sun as the main emitting unit of energy, used and transformed by the totality of stellar bodies, existing in its sphere of influence. In this way, the Earth receives approximately a force impact 1000 times greater than that necessary for the operation of all its physical systems. The planetary processing of light illustrates a universal principle in the treatment of energy; it absorbs a part that effectively transforms into work, while giving off the excess amount in the form of heat to the dark body that surrounds it, to the much colder space, in which it completely dissipates.

The observation of this principle of universal treatment of energy, by any type of system, led Nicolas Carnot to propose the first two laws of thermodynamics, a definition that perfectly reflects his sense and approach, that is, energy equals heat in motion. These two laws have become a true paradigm, probably the best valued by the scientific community and consequently applied to virtually all physical and social disciplines. The first law, which is taught in mid-level studies, establishes energy conservation. The total energy quantity existing in the universe remains constant, although as observed, it is transformed into useful work or use and heat or waste, called entropy. The second law states that any type of system, belonging to any category -mechanical, physical or biological- produces entropy. The constituent characteristics of a particular system, and especially its degree of complexity and specialization, determines the use it can make with the available level of entropy.

The first attempt to incorporate information and intentionality into the laws of thermodynamics came in the mid-19th century, when Scottish James Clerk Maxwell proposed a new scientist called statistical mechanics, in an attempt to explain of reality at the atomic and molecular levels. With the intention of combating the generality attributed to the laws of thermodynamics, Maxwell proposed a paradox, consisting in the possibility of separation of hot and particle colds from a gas, introduced into a watertight compartment. Divided into two compartments, by an imaginary being that had the ability to distinguish both types of particles and opening a separation door, all without producing friction or added energy. Physicist William Thomson called the figure a demon that can see every individual molecule in the box, just as he observes people's actions, to Maxwell's dismay.

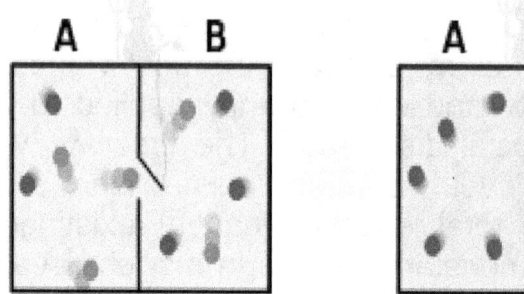

Maxwell´s demon represents the best-know paradox regarding the handling of entropy without generating additional energy. The dilemma is not true. In that any processing of information involves some kind of energy necessary for its production, wether electromagnetic or biological. Source: es.wikipedia.org

Clearly, the statements set out in the abstract analogy are not true. The storage of information, necessary to differentiate the two types of particles and the same exercise required for recognition, imply a rigged consumption of energy in the hypothetical and sympathetic demon.

The interest shown by science, in the treatment of entropy, may seem excessive to the reader, but it is the clarification of a transcendental fact in the configuration of the current universe and by extension, in the appearance of life on Earth; equally, in the possibility of life on other planets.

The second law of thermodynamics predicts that the most likely state of a system, regardless of its size or class, should be the state governed by the greatest possible disorder. This law allows a simple verification of its veracity, using a simple deck of cards. From its fully ordered initial state, at the time of opening the box, if you proceed to throw the cards into the air for an indeterminate time, the level of disorder is progressively increased, that is, chance determines the configuration of the system, ending in a state that represents the highest degree of clutter possible.

Likewise, the people responsible for cleaning the home have repeatedly found that dirt begins to appear, even during the cleaning process. Both configurations -the cards and the house- belong to the closed systems class. The evolution of life has taken place in an open system -planet Earth-, in which organisms in constant interaction with the immediate physical environment have shown the ability to transform reigning entropy into progress-oriented energy mainly aimed at survival. This fact defies head-on the law of entropy that governs universal physical behavior. The theoretical response to this seemingly contradictory phenomenon was provided by the Nobel Prize in Physics and Biology- obtained in different years -Ilya Prigogine, through the concept of dissipative structures; a configuration capable of transforming the resources of the environment into energy and expelling the entropy generated in that process to its immediate environment.

The human being also possesses the capacity for cell renewal but the amount of energy available for this process is limited. As noted above, any copying process incorporates errors; this is the case of replicating the genetic code in the cells. As the number of copy processes increases, another of the specific features of complex systems, namely the loop or vicious circle, comes into play; the agency must dedicate every time a new renewing episode takes place, a greater amount of energy to repairing the accumulated errors in the total of the replicating chapters executed. This dual channel of energy expenditure, ends in the production of cellular duplications too defective for its proper functioning; as a result, organic death occurs.

The empirical evidence available seems to confirm this fact. Cultured human cells are able to replicate in a cycle no older than 40 to 60 times -Hayflick limit-, at that point the process stops and the cell becomes senescent, descending its degree of effectiveness. This is the main reason for the onset in old age of degenerative diseases; the decline in the effectiveness of the immune system, more specifically t-celling. Human organisms cannot overcome -with the exception of exceptions- the 100-year barrier by producing cellular duplicates.

The reason for the death contributed by the information theory, therefore, lies, in a functional and not really biological flaw understood in the strict sense, but the similarity of this statement, with the proposal of indefinite existence offered by technology, resides solely in the diagnosis. The consequent extrapolation establishing a mechanical identity, between the improvement of the process of cell rejuvenation through specific techniques, such as the implantation of stem cells and the disappearance of the temporal limits of life, remains absolutely fictional. Non-mathematical estimates, handled in cases of cell renewal, range about a 100% increase. That is, 150 years -twice the current life expectancy- would be an acceptable estimate.

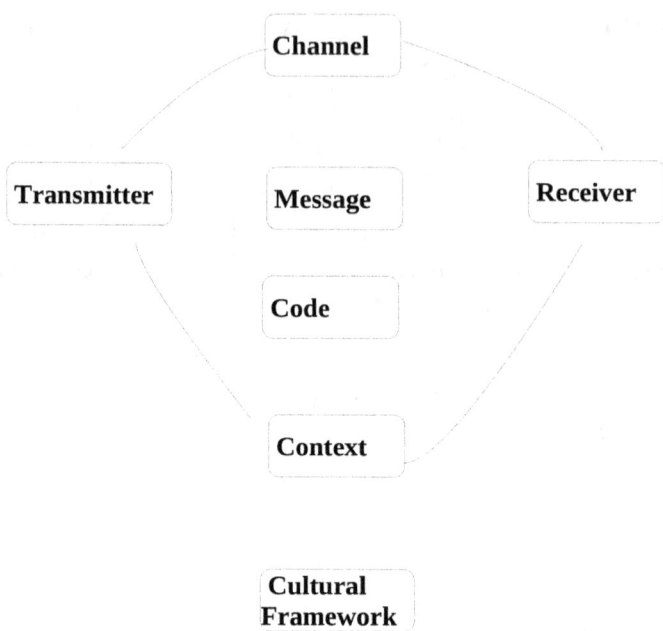

Information Theory assumes systemic budgets of interaction between its main components. Messages from the sender require a physical channel for transmission as well a code for expression and decoding. The are based on a common cultural context that provides a great framework for the interpretation of their meanings wich can be multiple. Source: es.wikipedia.org

Chapter XI. Forbidden Territories

"A science, that is, a knowledge imaginary of absolute truth."

Leon Tolstoy -Russian novelist-

"Science rivals mythology in miracles."

Ralph Waldo Emerson -American writer and philosopher-

A television documentary broadcast by a well-known advertising channel, presented genetic editing as the possibility of creation in the future, of little less than any kind of attributes desired in the human being. While an unknown commentator was vatileting about the achievement of an ancient human dream of flying like angels, a fashion specialist rambled about the possibility of designing skin color or generating any of the bodily attributes ambitions, thus achieving the conversion of the individual into a fashionable design in itself. The evocation of the scene of the saga "The Star Wars", staging a galactic tavern with a mixture of fauna of different species, seemed to derive from these proposals with the only assumption of the simple temporary pass.

It creates an uncomfortable feeling, contemplating a continuing series of false, as well as insidious, claims coming mostly from mass media. It is not possible to come to understand, the reasons for such a denient spectacle without the firm conviction, that this tendency represents an attempt to generate noise and debate, on a specific scientific caption, exhibited with frivolity, may resemble an advertising short film; and indeed, that's the way it seems to be.

Natural evolution, has taken at least 3700 million years, to travel a long path presided over by the simple but constant rules, regents in the evolution of living beings. Still contemplated this process with a certain coldness and estrangement, it still causes a sense of wonder and some wonder. Sexual reproduction, appearing approximately 1.125 million years ago and night vision, is estimated over a period of about 700 million years ago. All this, in a constant progression towards adaptation, with the aim of ensuring survival. It seems, therefore, a somewhat banal and overbearing attitude, to assume that even with specific techniques, a code as structured and specific as the human genome can be modified.

The possibilities of gene editing cannot violate the laws of natural evolution that have taken 3.5 billion years to consolidate. Dreams, desires and fictions cannot count on science to be realized are the object of fiction in many any its story formats. Source: Pixabay.com

The general confusion around this particular point seems to be based on a deep origin and is, of course, accompanied by serious ignorance about the rules of biological functioning in general and physiological in particular. Certainly, some existing resources today, such as the chances of cellular regeneration, can result in a significant increase in average lifespan. However, certain claims that seem to have become genuine modern myths should be rebutted.

It is obvious that, as noted above, organ implants from custom cultures, from the individual cells themselves contribute to extend life. Exactly the same effect, they produce cell regeneration techniques, especially stem cell grafting. But in no way do these applications lead to the yearning for access to eternal life. The inherent paradox is that the same mechanism necessary for the maintenance of life, becomes the main agent of cell oxidation and therefore biological death. Free radicals, are atoms or groups of atoms that in their composition, have an electron that is not associated so they are highly reactive and unstable. To establish balance, this atom will seek to "steal" an electron from another unit. When this happens, the atom that loses its electron becomes in turn a free radical. Thus, a chain reaction, called oxidative stress, is progressively generated, through which cells are damaged, producing widespread aging throughout the body. This continuous wear takes on the leading role in the onset of many diseases; the known elderly scourges. The remedy in an average lifespan falls on the regenerating role of antioxidants. The oxidative process, becomes a functional alteration.

External agents such as air pollution or tobacco smoke contribute to the action of cell deterioration, i.e. they increase stress. The really negative aspect is twofold. On the one hand, free radicals are not generated in the body spontaneously during the period of vital fullness but the phenomenon makes its appearance as the organism ages. In addition to this, there is a progressive imbalance between its production and elimination, the decisive cause of the disease and, as a direct consequence, of death. In particular, a decisive part of the deterioration is due to the continuous disappearance of the cells that direct the internal immune system, the T cells. When these useless cells die, they release dioxins -harmful chemical elements- that accumulate directly, in muscles and bones; in this way, a new dual negative state is created; loss of defenses and production of harmful substances. A very common interactive process, present in other degenerative phenomena of nature such as planetary overheating, with its redundant effect on rain reduction and desertification, interacting with each other. A clear example of a vicious circle.

Harman established in 1956, that aging was due to the oxidizing action of free radicals. Therefore, antioxidants can be administered to the individual, in order to lessen the effects of organic aging. Approximately 2% of the oxygen used by cells is not converted into water but into reactive oxygen varieties. Most of these species originate in the cellular mitochondria. Hence the importance of this and especially mitochondrial DNA for understanding the aging process. Like all complex functioning, aging is defined as a multifactorial process.

However, experimentation in the line of the discovery of the miracle pill persists. The mIT magazine, The Massachusetts Institute of Technology, published earlier this year, the discovery of one of today's most promising medicines, in order to combat aging, the product of a long and tortuous history. In 1999, the US. Food and Drug Administration approved rapamycin as an immunosuppressant -a chemical principle that attenuates the biological defense effects against external agents- aimed at rejecting transplanted organs. Scientists later discovered that it affected all kinds of biological processes; the "mammal target of rapamycin" (mTOR) increases immune function and decreases inflammation of transplanted organs.

The experiments also showed that rapamacin extended the lifespan of yeast, worms and mice. The question that came up automatically was inevitable: Could you do the same in humans? The drug targets people aged 65 and over, potentially increasing the body's immune response, to fight respiratory tract infections, the seventh leading cause of death in older people. Final results should be presented by the middle of next year 2020. Currently, there is no rigorous way to evaluate the potential of rapamycin, to delay human aging. Rather, researchers have focused on a significant aspect of aging, decreasing immune function, to check whether those drugs that mimic the effects of rapamycin, can improve immune function in older people.

The role of artificial intelligence applications, such as big data management and artificial simulations to scientific research fields, has been highlighted before. But due to the human tendency to unjustified extrapolation, projections linked to the possibilities of artificial intelligence are continually appearing. The most fantastic and repeated option by the mass media, as well as in debates and articles of all kinds, is probably the possibility of creating intelligent systems that can be refined autonomously and even more, that are capable of generating other cyber netting drives; that is, it raises the possibility of reproduction, similar to the biological cycle. Without having to detail the methods of cyberlearning, which can be produced in a way more or less assisted by external human intervention, one of the most meanings authors of the cyber discipline, William Ross Ashby, has been forced to reject the concept of self-organization explicitly and publicly, through which a machine or biological organism could change its own organization or in terms of the author, its "functional mapping". This expert, claiming that imagining this possibility, turned the particular process into "a bet of a metaphysical nature".

The primary condition for any type of organism, whether biological, or cybernetic, to acquire the potential or appearance of self-organization, requires the existence of at least one external factor that influences the aforementioned system. In the case of humans, this factor has emerged from the permanent environmental pressure and the continuous interaction between the two axes; biological evolution and physical environment.

The reductionist conception of man understood as a machine, has favored such proposals and again, computational simulations, especially in the mode of genetic algorithms - a type of programming that uses the laws of transmission hereditary to emulate a future image- have been able to show the complex perspective of higher organisms and especially of the human being. Indeed, Descartes and LeMettrie, among others, proposed the theory of bodily functioning, understood as a mechanical body, thus substantiating the belief in the existence of spare parts, as a path to an indefinite life cycle. A set of materialistic beliefs extended the analogy between universe, machine and man. The maxim "Deus ex machina", original of the Greek and Roman playwrights, would later be reinforced by the discovery of the great universal laws of cosmic functioning and also by the second industrial revolution with generalization of the consequent machinism. The shocking contribution of the law of the "continuous movement" represented by the "pendulum of Foucault", also calls for its decisive role in this conception of nature. Nietzsche and his famous phrase "God is dead", has subsequently been used to reaffirm the desired law of the pendulum of perpetual movement, as well as the liberal meta-theory of unlimited progress, establishing a decisive turning point in the conception of human becoming.

A central issue underlying the discussion, about the possible indefinite continuity of the body, rests on the central aspect of the analysis of complex systems, one of whose prototypical cases is the human being. Specifically, the role of chance or determinism, as Jacques Monod expressed in 1971, in his work entitled "Random or Necessity". In it, the author opted for chance as the fundamental and decisive determinant of human evolution. Contrary to the randomist ideas of Bateson, Monod and Morin, John Holland's genetic algorithms and John Koza's evolutionary programming have shown that in a growing number of models, the role of noise, mutation and accident, i.e. of elements that generate randomness or chance play a completely marginal role.

Comparison with the capabilities of algorithmic operators, such as the recombination of individuals -cross-over- and natural selection, championed by the most numerous thinkers and scientists, among which the double Nobel laureate Ilya Prigogine and her concept of dissipative structures, as a generating element of order in biological organisms, have shown a decisive role as guiding principles of evolutionary becoming. All these theoretical principles have shown that complex organisms adaptive to the brink of balance are fundamentally governed by rules presided over by order and internal coherence, even maintaining the constituent characteristic of unpredictability, which defines precisely complex structures. That is, it is not possible to predict the likely behavior of a dynamic complex system, with total accuracy. All knowledge is clearly probabilistic, for this reason, the totality of physical experimentation, begins the communication of its results with the well-known phrase "Under normal conditions...".

Nor is it supported today by Morin's belief -strongly criticized by some epistemologists as Maldonado- that, chance acquires the greatest prominence in the behavior of complex systems; this is especially true, in cases where the volume of time or space takes on large dimensions, such as biological life or social behaviour. In short, there are other predictive methods that are more efficient with respect to complex systems. The basic question that serves as a conclusion rests on the fact that systemic dynamics are more accurately explained by rules governed by principles of order and coherence, than by a set of random or spurious circumstances. Otherwise, science itself would become an impracticable activity and continued evidence demonstrates the contrary statement.

The most widely used pathway for progress in the development of artificial intelligence has been oriented towards the discovery of brain function. This is a task at least, arduous. First, software -computer programming- and hardware -the physical base or "iron" - perfectly distinguishable in a simple structure like a computer, are in constant interaction in the brain. Thus, the frontal dialog or discussion between structuralism and functionalism that in the middle of the last century, concerned Anglo-Saxon sociology and philosophy, especially American ones -if it ever exists or has ever existed- necessarily turns out irrelevant, since human beings and their abilities are the result of the obvious interaction between structure and function, especially in specific intelligent processes.

In addition to this and far from representing a still photo or a static frame, the brain structure and its functionalities, are in constant change and evolution, taking full sense the definition of the mind, as a structure with adaptive flexibility. In addition, the brain develops itself, during growth and even in adulthood, replacing connections and generating new neurons, in a considerable volume and with a frequency of high voltage. With a consumption of just 20 volts- just like a low-voltage homemade bulb -brain processing achieves performances similar to modern computers, and like them... in the most absolute silence.

However, human beings have little control over emotions, since one of their essential characteristics is the speed in which they arise and in the same way, the causation of their effects. The emotional origin is located, in the limbic system -a structure corresponding to the sympathetic or involuntary nervous system and shared with the upper mammals- and not in the prefrontal cortex of the brain, the basis of elaborate thinking. In other words, there is no rational control over the emotional response. Emotions are essential to achieve the balance of the cognitive system, in which sleep shows a prominent prominence. Over the course of rest periods, the subconscious regulates brain balance and therefore the stability of individual consciousness.

This is the main reason why a large number of authors are discarding the idea of possible equality, between biological intelligence and artificial intelligence. Clearly, the range of the force of rejection, as well as the arguments used, have a high variability, although not without reason. Common and important arguments lie not only in the inability to emulation the wealth of life, but fundamentally, in the absence or lack of essential parameters of a functional nature, essentially of a motivational or non-essential nature. Thus, defects with the greatest number of mentions, refer to the absence of distinctive features of human intelligence in its virtual counterpart, such as the will to live, desires, appetites or fears. In short, at best, artificially intelligent units only compute and learn within logical schemes, but lack the most characteristic element of human evolution: irrational pulses. In other words, an ancient hominid, endowed with a perfected artificial intelligence, could not have survived a single full day in the age of the dinosaurs, nor just, an attack by a rival tribe or any of the multiple cataclysms that have plagued the evolution of hominids on earth; one of the phrases that best defines the analogy between the two types of intelligence -biological and artificial- has been uttered on many occasions and places: "soulless brain".

Despite all this, the possibilities of technoscience are wide and significant; they do not need exaggeration or less, of repeated distorted. One of the constituent characteristics of new technologies lies in their confluence or otherwise, their joint use. Thus, the first remote surgical operations, using robotic assistants or not, have been performed using 5G communication technology that significantly reduces the latency time between communications. In Spain, Dr. De Lazy has carried out the first operation of these characteristics, located 4 km. away from the hospital Valle Ebrón. Similarly, when Dr. Mehran Anvari takes his surgical instrument and makes cuts to a person's body, he does not use his own hands. In fact, it's not even in the same room. It surgically intervenes, patients 400 km away. From a console at St. Joseph's Hospital in Hamilton, Canada, he controls a surgical robot, which is located in another part of the country, which cuts, sutures and removes parts of the body. So far, it has performed more than 20 operations, of various kinds such as colon interventions or hernia repairs.

Progress in different areas of technology has enabled the viability of such applications. Augmented reality - AR- and virtual reality -VR- are becoming capable and affordable technologies for widespread adoption. Telecommunications companies are deploying 5G networks at a fast enough pace to allow the handling of huge amounts of data, from data-based arrays supplied by advanced sensors and with no appreciable lag times. R & amp; D & amp; companies -Research, Development and Innovation- that work in this field, perfect technologies that allow different people to interact physically even if they are located in separate physical spaces. These techniques add -touch-to-touch- haptic sensors to your robotic avatars. The so-called total sensory immersion, an end point for the application of collaborative telepresence, will require a decrease in current delay times in video calls, although using a small ruse, the perception of delay by could be eliminated by predictive algorithms.

In short, a robot used in telepresence practices, can be considered as an extension to video calls, where the remote computer is located inside a mobile robot. This type of application, is very useful in emergency situations in which the doctor or surgeon is geographically distant from the hospital center and therefore the sick and above all, in circumstances of extreme gravity, in which time becomes a fundamental variable. The specialist with a call and/or video call from his mobile terminal, could have a detailed knowledge of the situation and provide directions to a telepresence robot, located in the hospital. The obvious condition is limited to the availability of mobile robots in the care centers. The doctor would begin to control the machine, starting with its movement to the specific room or point, where its presence is required. The cameras and microphones inside the android, exercise the function of eyes and ears, all this, as well as the screen would show the arrangement of nurses and helpers, all in remote mode. Interaction with the equipment and the patient would be facilitated by zooming and sound volume.

Identical telepresence benefits, show their usefulness also in the areas of work and education. Distance education or participation in conferences or seminars, no matter where they are physically held, is already a reality. Similarly, in research centers, their applicability proves highly effective and useful. The hologram image -the virtual reproduction of a typical sci-fi human figure- does not seem to be necessary anymore, due to this type of technology.

Artificial intelligence, shows in one of its most important specialties, robotics, all the influential elements in the social impact and consequent assessment, of a significant scientific advance. The areas of influence of robotics are multiple. The application that arouses the most attention, probably resides in its use for current work and the ghost generated by the replacement of human operators with robots, of great media impact. This fact has been assessed from its most negative perspective, as a social threat and therefore a tendency to counter.

This same discussion has been going on for almost 50 years, specifically in the automotive sector, one of the first to apply robotization to the automotive assembly chains. At the end of the massive use of the human operator is Ford, the Company of North American origin, which until relatively recently, has resisted to robotize its tyrannical assembly chains. In this company, the first studies of times and movements were carried out, with the aim of improving and standardizing the production methods. Under such actions, the ultimate goal was to increase the economic surplus. Unremissibly, she was forced to apply human motivation techniques -the school of human relations had her birth here- in the face of failure in the attempt to force a human being, to perform tasks unbecoming of his characteristics, both physical and mental.

Robotics offers a wide range of features in wide variety of human activities, from performing repetitive tasks to telepresence options in which audio and video devices are incorporated into the robot wich allow the intervention of professionals in different geographical locations, intervening or communicating with other specialists, as well as having a full view of the distant situation. Its applications cover the medical, educational and business areas. Source: Pixabay.com

At the opposite end, the Japanese company Toyota, considered for many years and today, the world's largest company in the sector. In its central factory, located on the island of Aichí (Northwest of Japan), it has never used human operators since its inception, except in supervision tasks. However, with part of its considerable economic surplus, it created a university for the community in which it operated before its inception. It should be emphasized that, however, the cultural shortsightedness prevailing in much of 21st century society continues to project fallacy motivations, in the face of a fact that shows a clear and widespread corporate social responsibility.

The extrapolations, in this case, are favored by the use of misleading associations, understanding that the advanced university training provided by this university, is unfailingly linked to the technological sciences. This reasoning is absolutely false. This fact offers important considerations and of profound significance. The first and fundamental, is that society as a whole must decide the priority objectives in the re-investment of the economic resources generated, from any source, both from private and corporate profits. This channeling is carried out through the fiscal policy enacted by any of the world's governments, but it does not exempt the various social actors from their obligatory compliance.

However, in recent consultations with solvency computer programmers, these professionals reveal a well-known data, which means that intelligent applications, even in their own programming activity, are only capable of engage in boring and repetitive activities. It is therefore appropriate to use it and not only that, but also welcome.

It is worth noting, in the assessments made about robotics, the persistent influence of an ancestral human tendency. It refers to antropormorphization or propensity to attribute human characteristics to non-human ideas or forms. This inclination probably cannot be completely disjointed from ancient religious animism, inclination to attribute life to real or imaginary entities such as sickness or imagined gods. Humanized robots and astrological figures, projected on stars and galaxies or simply, in the usual cloud formations, are clear examples of these ancient tendencies, which seem to permanently accompany human perception.

The result of the conjugation of these facts determines a strong demonization, not only of robotics, but of artificial intelligence as a whole, in many representative social sectors. It is appropriate here to highlight certain facts, which, because of ignorance or undue generalization, seem to be generally disregarded. The first, it derives from the use of common language -as elsewhere, could not be otherwise- to refer to or evaluate scientific facts or specialties. The use of metaphors, analogies, euphemisms and hyperboles, constantly dot the expressions of the lexicon to use, producing a confusion of decisive weight. This same fact affects other relevant human activities such as politics, which has been demanding specific language for some time, in order to avoid a miss-transmission of concepts and ideas to society. This type of language should be very similar to that used in formal logic, of type "if A is unequal of B and B unequal of C, then A cannot lead to C "-A/B; B/C>>A/C-. Unfortunately, this is not the case today; dialectical pirouettes used by politicians in general, regardless of their ideology or belonging, exempt from a genuine demand for formal coherence in the expression of ideas, which in most cases have major implications for effective social functioning.

The equivocal use of common language leads to absolutely inappropriate and non-scientific fact or data, even used by specialists, critics and media, such as the issue that acquires the greatest spurious and equivocal relevance, about artificial intelligence consisting of sentences similar to "machines that learn for themselves can come to dominate humanity in the near future and even destroy it." It is clearly a conceptual bias, determined more by beliefs and prejudices than by objective evidence. The errors, rational fallacies and false claims are so many and so varied that they make it difficult to make an orderly presentation difficult. On this particular assertion, sufficient arguments have already been provided that if repeated again would clearly be redundancy.

The fundamental physical principle is universal, in any system and with different time scales, every body that has a beginning is born with a programmed end, from the bacterium to the same solar system that will explode within 5 billion years . It is understandable that in the case of a being endowed with symbolic intelligence as the human being, it is difficult to accept its role as a mere link, even if it constitutes the being of greater degree of perfection in the whole of the vast cosmic ensemble.

Chapter XII. The Recurring Utopia

> "If I've seen beyond it's because I've risen on the shoulders of the giants."
>
> Isaac Newton -British Physicist-

There are two social axes, which constitute the most studied headings and bring together the largest number of publications, articles, books, interviews and documentaries in the current era. These are the labels of power and technology. The historical constant of alliance between the two makes them recurrent parameters, of indispensable presence, in the definition of successive groups and communities, which have formed throughout human evolution.

Power as it is now known anchors the beginning of its relevance in the religious conception of the concept. At a certain time and place of social evolution, it adopted its maximum expression and standardization. Egyptian civilization has acted as a catalyst and directional source of a format of power that is absolutely decisive in the human historical process.

The key concept is the supposed mechanism in the transmission of power, which historically has a divine origin, belief has persisted to this day. The acquisition of authority automatically by kings and later emperors was blessed by the absolutist medieval church. The State took temporary respite from the exercise of command, without in many cases softening the modes of government, as well as the pressure on citizens, both in the economic area, and on the compliance with the laws enacted and the necessary attitude of submission required for compliance. An ancient collaboration between the two types of command -religious and political- generally maintained by the different societies that have formed during the historical development. Derived directly from the stage in which the magical -mythical- knowledge was firmly installed in the popular imagination, astrology and more specifically, the central idea of predetermination, translated in vital terms incorporated acceptance of the previously written and pre-established destination, they ended up in the formation of a very specific syndrome in citizenship, catalogued and labeled by the social sciences in general, such as the helplessness syndrome learned.

Specifically, it consists of the firm and widespread popular belief that life, both in the individual and group spheres, are determined by external and spurious events and phenomena, which are therefore totally self-willing and determination of individuals and groups. This phenomenon is described and incorporated into the theory of attribution, which acquired considerable relevance and significance in social psychology. Contrary to what could be thought, this attitudinal slab persists to the point of categorically defining totally unrealistic sources that crystallize in the widespread submission of complete social groups. The trend has been increasing until the end of the last century, without its end point being observed, typical behavior, on the other hand, in the processes that govern the appearance and behavior of fashions and idioms, of general influence in any society.

We have coined the term parental home syndrome, to refer to the renunciation of social groups from making self-determining decisions, which govern the fate of their own destiny. It does not obviously refer to individual decisions that have a reduced influence on the lives of subjects, such as profession, love or work, but, and fundamentally, the depth and direction that must be taken by the behaviour of the societies of the Future. The persistent human tendency towards accommodating and maintaining past beliefs and traditions, such as the determining weight of history or the existence of a written destiny, should be conceived as a genuine negative agent while contaminating in the social health and an inhibitor of true progress. Of course, it is a dynamizing element of systemic conditions, such as inequality and the loss of hope in the time to come; in essence, a barrier to the overall motivation to face the challenges of the future.

There are basically two categories of mental components, which affect effective behavior or behavior, in this case at the social level. The first is formed by the perception and interpretation of the facts, above all, those that have a marked historical impact. This first axis, affects history as the determining axis of the present situation and consequently, of the options available for the future. Indeed, the past generally contemplated from a historical perspective, that is, as a sequence of concatenated and consequential facts, leads directly to the fallacious conclusion, that the defining limits of the present, are of a character decisive and immovable and worse, necessary. The general reasoning is that if history has led society to this point, that was the case.

The second group of elements refers to ideal constructions or virtual constructs, on which the false belief of social determinism, which are present in the collective idealism, is based. The imagined stereotypes have set up an authentic constructed reality. Thus, until the beginning of the last century, the occurrences of religious tone constituted a broad demonstration of the existence of divine principles. Testimonies of unexpected apparitions and religious messages to ordinary individuals multiplied in the early years of the last century. Subsequently, the type of revelations suffered a progressive slide towards UFO sightings and abduction accounts. All of them, uniform, repetitive and following a generic scheme, guided and prototypical, common to the whole of the reported case.

It is up to an incredible kind, that even scientifically leading states, such as those formed by scientists and researchers, repeatedly and as a supposed condition, the causality of sightings of flying objects does not (OVNIS), to the presence of one or more extraterrestrial civilizations that have reached the planet. This reasoning actually represents a very common practice in the production of false theories. The procedure is simple. First, a theory is built based on unproven facts, in this case, the existence of alien races. Logically, these kinds of theories contradict official science. At the moment, when scientific sources for use, show their inability to produce a full explanation of a particular phenomenon, usually for lack of valid data, their causality is attributed automatically, to alternative theory, not counting in no case with demonstrative facts, beyond mere conjecture.

But the fact that he wishes to emphasize, consists in the award of a purported and in fact, in need, technological superiority of extraterrestrial civilizations. This phenomenon clearly demonstrates the habit generated by historical and sociological traces, of continuing the search for reference patterns that offer the right path of behavior to follow. In other words, humanity still needs an "older brother" similar to the direction of parents, who determine behavior and assume responsibility for popular action, especially that which refers to the correct course of future behaviors. It is not denied from here that the primary motivation of NASA's extraterrestrial intelligence search program -SETI- primarily obeys this goal, however, some of its basic assumptions include as previously stated, the seeking alien civilizations for a supposed level of superior technological and cultural development, with a higher development rate capable of providing a behavior guide, a correct vital decision-making process. In conclusion, the transfer of liability and the right of self-determination to another external and above all, superior agent, highlighting in the above statement, the qualifier of "right decisions".

Throughout the history of human development, the moral guidance of the populations has come from the outside of their own reason. In ancient times, the Lighthouse of Alexandria was a reference of all known, for the orientation of sailors from any point of civilization. It was a point of orientation, which once sighted, in fact presupposed the possibility of fixing any desired maritime course. Similarly, in the adolescence phase, at the time of a problem, considered serious for the individual, the usual friends asked the decisive question, about the potential resolution of the conflict presented. The question was "do your parents know?" or "what have your parents told you?" The actual state of the problem and its solution was going through the hard time informing the parents, insofar as they meant the principle of total authority. Once the moment of pressure had passed through, the situation causing the problem could be considered solved, as the resources of the parents were considered virtually infinite.

The search for the lighthouse of Alexandria for the orientation and syndrome of the house of the parents, acquire an identical meaning. They represent at the same time, the success of the decision to be made and the evasion of individual responsibility with the consequent psychological relief that this entails. To paraphrase old Aristotle "fear is the feeling produced by the expectation of an evil". Once the guide for your solution is found, all fear and anxiety disappear and very often... magically.

Divine Power

Religious, military, administrative Power

Writters, technicians, whealers

Workers, free mans

Servants, slaves

Egyptian civilization built the great pyramids, burial monuments, actually representative of the belief in the mechanism of transmission of divine power in all social orders, not for nothing, the Egyptians were the greatest defenders of the division of labor and social castes. Source: Own elaboration.

Chapter XIII. The City of Stars

"Everything is theoretically impossible, until it is done."

Robert A. Heinlein -American fiction writer-

"We are an advanced race of monkeys on a planet minor on a very average star. But we can understand the Universe. That makes us very special."

Stephen Hawking -British theoretical physicist -

The second half of this century will witness, no wonder, the first manned mission to the planet Mars. The mission has a permanent character, constituting the germ of the future human colony on the planet. Two space agencies -the US space agency and the European space agency- together with two governments- Russia and China- together with a private company -Space X- led by Elon Musk, an entrepreneur of various business initiatives, have united their efforts to launch the future mission.

The official justification is understandable. It is intended to diversify the unforeseen risk, any imponderable threat that may affect humanity decisively. With this project, humanity will become a bi-planetary civilization through this dangerous alternative, for the first time in its history. However, a variety of concrete real reasons can be argued, supporting the completion of an attempt with obvious risks. First, the most significant consequences of the prevailing capitalist model, from the end of World War II to the present day, can be classified as serious and probably irreversible if taken as benchmarks, the model of predominant society and the environment.

Climate change, as a result of global warming, will in the immediate future cause serious damage in the near future, which is irreversible. Without the need to admit Gaia's theory -the consideration of the planet as an entity endowed with its own intelligence- the planetary response with high rates of pollution and overheating, due to extractive, consumerist and typical of neoliberal ideology, presents a range of unsettling probabilities.

The most important consequence possibly focuses on the poisoning of the three basic ecosystems, water, air and land. Levels of breathable air pollution clearly show the inability of governments to implement emission reduction measures, agreed at all the climate summits that have taken place since the end of the last century so far. The interests of large productivist corporations continue to dominate global overall well-being. With regard to landfills into the oceans, not only but mainly industrially responsible, it is worth highlighting the heavy elements - such as mercury - and plastics that have already contaminated fish and mammals. This is not just an ecological alert. At the moment, fish intake has become a real risk of food poisoning, for the common consumer.

Global warming carries the risk of the continued trend of defrosting, polar ice caps and the transfer of these waters to the oceans, which will inevitably increase the level of their waters. If during this century, the planetary temperature continues to rise, reaching the critical point of 2 degrees, the sea level will rise by a minimum of 50 cms., starting the global flood of a large number of coastal cities.

Regarding the soil system, similar but probably aggravated circumstances are observed. The increase in the degree of air pollution and the progressive drought due to the alteration of the water cycle will culminate in a general desertification, which has already begun. In addition to this, the indiscriminate logging of trees in the main forested areas of the planet, which act as global lungs, as is the case of the Amazon rain-forest, accelerate the process.

As if that is not enough, the danger of the glaciation cycle is also affected. A glaciation is a long period, in which global temperature drops and the result leads to an expansion of continental ice towards polar ice caps and glaciers. Drastic freezing is subdivided into periods, with the last recorded to this day. According to the definition, the term refers to a period with the formation of glacial ice caps -both in the northern and southern hemispheres-. In more colloquial terms, when talking about the last millions of years, the name "glaciation" is used to refer to colder periods with extensive glacial caps in North America and Eurasia. According to this definition, the most recent cold wave ended 12000 years ago; the next, under normal conditions, should occur over a maximum period of 30.000 years.

The future human colony on Mars will suffer the initial pressure produced by the harsh physical conditions prevailing. On a planet where the atmosphere represents only 1% of that existing on Earth, with a gravity of 1/3, compared to the existing values on the original mother planet and a daytime temperature that can reach 60 or Celsius, the only real option consists of life under protective structures of materials, specially developed for this purpose, for sure, some type of polymer. The degree of technological development available to the mission will be of a high level. There are already telematic and inflatable structures, whose use and installation is carried out almost instantly. Once installed, its first objective will focus on climate change, i.e. the beginning of the transforming process, the primary purpose of which is oriented to global warming, which would enable the defrosting of the permafrost or ice layers existing on the ground and in polar ice caps. It is possible that within approximately 200 years, the planet will evolve into the desired green phase, with the appearance of flora transported from the source planet.

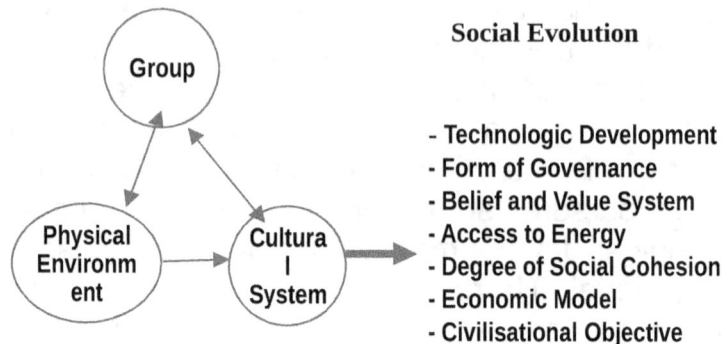

Life on Mars will developed conditioned by determinism due to adverse conditions arising from the psychical environment. There will be as has already happened on Earth a strong influence between the environment, the group characteristics and the cultural background of the same group certainly determining a new type of evolution. Source: Own elaboration.

The new colony will undergo changes in virtually all of the cultural components, which will form the initial group. A typical case will be observed in the form of community governance. In a first phase, the rules from the mother planet will be reproduced with strict fidelity. A joint command, consisting of a scientific officer and the inevitable military officer, is foreseeable. In time, it will lead to a participatory democracy, overcoming the usual deception offered by representative democracy. All this, before the stage of receiving settlers begins. It is estimated that with a size of 125,000 members, all the functions essential for Community functioning would be covered.

The colonization of Mars represents both and adaptative milestone as well as an opportunity for a new social design. Until you open the settler admission process, adaptative stress will be a considerable effort. Source: Pixabay.com

The entire process of colonization will involve considerable effort, at the top of which, the forecast of continuous adaptive stress is considered colossal. This parameter will act as an undoubted enhancer of the degree of group cohesion, thus facilitating the management of essential processes for group establishment, such as decision-making and governance in general.

Every advanced civilization depends on a direct function of the energy supply; and not only this concept incorporates physical energy. In this section, solar energy captured by photovoltaic plates will provide a more than sufficient level of supply, as in fact, done on Earth. With regard to motivational and relational stimulation, which feeds any group, the superlative objective of survival, will serve as a central parameter for the correct group functioning.

One of the civilizing components, which will experience a greater convulsion, refers to the inevitable economic model, prevailing in any type of civilization. For the first time in human history, a community will effortlessly access to the coverage of its primary or physiological needs, as well as superior ones, including effective, relational and personal self-realization. In the first group, food will be grown in greenhouses, using techniques, already well experienced on the mother planet. Hydroponic crops, together with the practice of growing transgenic varieties, will satisfactorily address the need for which all terrestrial animals have fought for millions of years. The dress will be custom made and produced by 3D printers, with insulating materials and safely, self-protective. Medical demands will be significantly reduced, relative to terrestrial patterns. The devices indicator of the basic vital functions will be for general use. In extraordinary cases of urgency, the generation of components and organs manufactured from fully customized genetic storage would solve any contingency in this area.

Information and stimulation requirements will be obtained through intranet connections and virtual reality-based techniques, which will serve as highly improved means, in order to provide the necessary periods of relaxation and fun, compensatory of hard work and a life, conditioned by continuous extreme experiences. Thus, without needs that require coverage, another of the evolutionary axes of terrestrial societies will be meaningless in the new civilization. There will be no exchange and therefore trade derived from it. The great conventions and functional agreements of the Earth planet will disappear, along with its associated institutions. There will be no place for the use of money and so, large production or financial corporations will not exist. Nor modern myths such as bonds, bonds, financial derivatives or debt.

The expected end of the monetized economy, and thus, of financial capitalism as well as productivist and extractive practice, will take place in the presence of a privileged group of witnesses. At this point, it should be noted that the search and extraction of new materials, as well as those known, will not stop but will not be carried out with a specifically commercial intention. Indeed, the most important treasure, knowledge and technique, will have traveled with the group. The children and grandchildren of technicians and engineers, with a high level of intellectual and witnesses to technological progress, the belief system will undergo inevitable changes. As is currently happening on Earth, religious beliefs are at a clear and declining goal, especially in the group of subjects under the age of 40.

Used to the firm mental discipline of objective data, there will be little room for extraordinary and abstract mystique. Paying attention to visionaries and speculators, he has overemphasized the changes that the bodies of resident humans will undergo. It seems clear that the decrease in severity, will result in a lower muscle and bone mass, as well as a greater height. The body regulatory system would automatically allocate more energy resources to the brain in that case. Similarly, it has been stated that the absence of contact with the outdoor environment would end up with a loss of body hair. The stereotype of the alien of considerable height and thinness, with an enlarged head and eyes more.

Because of the lack of natural light flow, can be admitted as probable. Time removes and gives reasons; some born in the present century may witness a new uniqueness. Finally, in a stable, equitable and satisfied society, a civilisational purpose, a civilisational fabric, must occupy the primary objective range. It is understood that this can be none other than to continue space exploration starting from a privileged platform, located more than 150 million kilometers from the mother planet. If these assumptions of stability are met, a central question would automatically arise: Does the long-standing achievement of an equitable society represent an illusion? The opposite option has marked social evolution on Earth, managing to implement a constant rule, encrypted in the continuous attempt at animal origin, directed to the domination of majorities by the power elites. But if the idyllic representation of a just and equitable society were feasible, then how long will it be until the new colony becomes a self-sufficient and autonomous cell and demands its independence from the old civilization Terrestrial?.

Chapter XIV. The Navigator's Flight Eternal

> -I'm sure the universe is full intelligent life. It's just that it's too smart to come here.
>
> -Arthur C. Clarke - British writer and scientist -

The universe is governed by four types of forces, which act permanently but with different scopes or spheres of application. If observed upwards, they govern behaviors from the atom -the atomic level- to the cosmos -macro-physical level-. The weak and strong nuclear forces operate at the atomic level. They represent the cosmological configuration in its early stages, constituting the regulatory basis of the only macro-structure in which life could have arisen and at a later time, intelligent life. The first -weak atomic- holds the electrons attached to the atomic nucleus. The strong variant, guarantees the union between different elemental units to form the molecules, that is, the union between different atoms, belonging to the same element -oxygen, nitrogen, carbon- or between different elements such as water -H_2O-. Electromagnetic force operates locally or planetary. It allows the lighting of cities, as well as the development of communication technologies such as the telegraph, the telephone or the specifics of medical research such as scanning.

Finally, in the universe and large structures like planets the gravitational force predominates. All terrestrial physical latticework, especially living organisms, has developed under the influence of gravity. This force conditions in the case of superior mammals, each of their bodily characteristics and functions. Determination of bone and muscle structure, as well as blood circulation and breathing. The pumping of the heart with the phases of systole and diastole -contraction and expansion-, can occur by virtue of the weight of the tremendous volume of air existing on the heads of the individuals, attracted down by the gravitational pull penetrating by the nose and mouth , facilitating the thrust from the inside of the body to the outside. The same case is observed in pulmonary respiration, through the movements of inspiration and expiration. Blood circulation, indispensable for the cellular supply of oxygen and food, is possible thanks to the gravity that makes it possible to circulate. It is clear that the anchoring of the human body to the gravitational constant is seriously affected at the time of abandoning its influence, as is the case of interplanetary travel regardless of its duration.

The four classic stages of any space journey are usually classified as first and last phase and do not usually present, unless unforeseen, greater difficulties. Dangerous complications and key limitations are found in the intermediate stages, -i.e. in the navigation and communication phases-. In the navigation phase, radiation in deep space takes on such intense activity that both the ship and astronauts must be protected. Radiation may be more intense, especially in the presence of cosmic rays and solar flares. In addition to this, the danger caused by the bombardment of meteorites, especially intense in deep space, should be added.

The most suitable solution seems to be in the coating of the ship's structure, by means of a shield of identical characteristics as the aggressor source; a cover or shield of hard material -probably graphene- electromagnetically charged. After spending a few months in weightlessness, the body loses a significant proportion of calcium and other minerals, causing astronauts extreme weakness, even when performing daily physical exercise. It is estimated that some of these effects of loss of bone and muscle mass can be prolonged permanently, so an astronaut will feel the consequences of prolonged weightlessness for the rest of his life. Acute irradiation syndrome, and even cancer, may be some of the added undesirable referrals.

All of these effects would be inoperative in the event that a robotic navigator was used as a substitute for the human crew member. At the time of imagining this situation, it tends to think of a robot morphologically similar in proportions and appearances to a person, no doubt by the general tendency to confer a humanized image on objects, whether known or not -Anthropoformization-, but the most likely option approaches the robotic form of the entire entire ship, led by a complete mechanism based on artificial intelligence's own programming. Thus, space exploration in search of new planets, for different purposes, would continue indefinitely. In the words of Dr. Nicoelis "We will probably be able to remotely control our emissaries and ambassadors, robots and aircraft of many shapes and sizes, which we will send in our place to explore other planets in the far reaches of the universe..."

With regard to the communication stage, there is also no noteworthy difficulty. Work is currently being done with the possibility of incorporating information units into the laser beams, but certainly, the radio waves reach a speed close to that of light, as long as they are emitted and received by a computer with sufficient processing capacity.

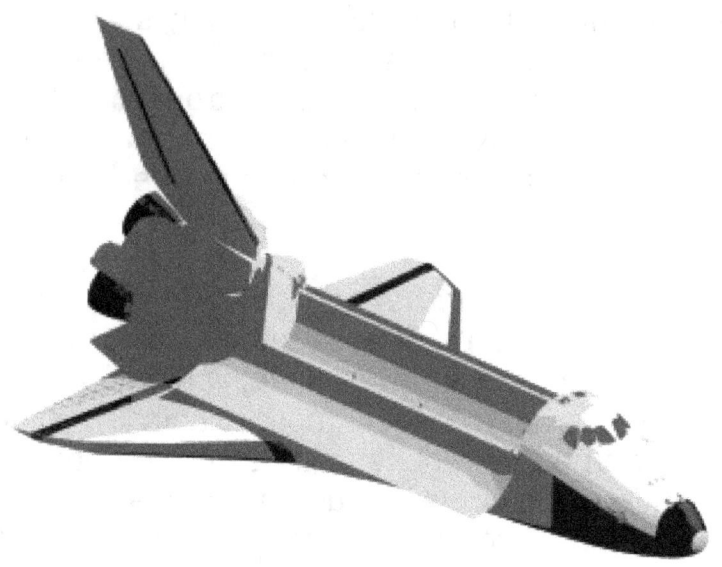

The possibility of replacing human crew members in future space travel with robotic entities is getting closer and closer to reality. The disappearance of existing threats to individuals together with the decrease of risk of permanent loss and injury allows to glimpse the possibility of prolonged navigation without identifiable end. Source: Own elaboration.

The other relevant section of navigation refers to the type of fuel used closely related to the potential speed and therefore to the duration of the journey. There are several totally viable possibilities here, which are already being experienced. The first consists of the use of plasma -plasma derives from the heating of a gas, which in that process is ionized, that is to say that it is electrically charged-. The whole universe in practice is made up of plasma. In the same situation of abundance, there is the use of hydrogen-powered engines -the most common component of the universe- and engines moved by this element have long existed. Navigation using drop-down candles provides a newer concept. Space probes or spacecraft would deploy huge sails -of the height equivalent of a 10-story building- that would receive any type of energy available, in the specific location where the ship is located, mainly of light, that is, photovoltaics.

As is already a custom, in almost any section concerning future technology, inappropriate derivations of scientific principles arise. On this occasion, there is speculation about the possibility of "doubling" space-time. This option is based on one of the statements in Einstein's theory of relativity. Certainly, the hypothesis considers the space-time continuum, as a geometric dimension and therefore modifiable... but only in theory. The amount of energy needed to achieve this effect would exceed in billions of times, the very energy dimension of the universe as a whole. There could be no shortage here, another derivation of the theory of relativity, in particular the reference to the deformities of the tissue that forms the constitutive matter of the cosmos. The existence of black holes and the so-called wormholes -which supposedly connect two non-continuous areas of the space surface- emerge as travel alternatives. A wormhole has never been observed; their topographical characteristics, as well as their size, location and inner workings are unknown. For now, the ship in the film "Star Trek" -Enterprise- will continue as a desirable and spectacular model restricted to fiction.

While all this is happening, the search for new planets, fit for the development of life, continues. The first significant data about the observation of the known universe was obtained through the Hubble Observatory, located on Mount Wilson, California, during the 1960s. The next group of relevant inspections is attributed to NASA's Kepler mission. In the search for outer planets that can host life - current or future - attention is focused on the existence of any of the three conditions, considered fundamental to the emergence of complex life. These are the existence of water, an adequate temperature range, called the habitability zone and the presence of simple active ingredients. That is, it seeks to detect planets orbiting active stars, which meet any of these three conditions.

The Kepler telescope has found more than 4.000 planets that meet those requirements. Apparently, the number of stars with planetary systems orbiting around it far exceeds the estimates initially made. Kepler by running out of fuel probably in 2018 will have ceased to exist. Their release will be taken over by another satellite, this time designed and manufactured by the European Space Agency, in which there have been high hopes of finding unanticipated results, less surprising regarding the number and adequacy of the conditions necessary to house life. Kepler has discovered potentially habitable planets, just 400 and 1200 light-years from Earth. The greatest impact for astrophysicists is the so-called super-earth, a large planet orbiting the Barnard star and located just 6 light years from Earth.

The detection of exoplanets is carried out using different observation techniques. The most commonly used handles gravitational micro-lenses, such as those used by the SETI program; it works at the exact moment when the gravitational fields of both the star and the planet overlap, thus enabling the location of a distant star. In general, all of them are based on the effects produced either by the light or gravity of one body on another, so that whenever it is possible to find it if direct observation is used.

The search for exoplanets with the necessary conditions for the development of life is motivated by two main causes. First and foremost, because it would constitute the undisputed proof of panspermia theory, which maintains the basic principle that given the right conditions, life is a widespread process throughout the universe. Any discussion of the metaphysical option represented by creationist belief would end radically. Second, there would be another viable option for human colonization besides Mars. A third previously indicated cause, consisting of the permanent search for a guide, facilitated by any type of top-level entity, in this case, of a higher intelligence of alien origin, should be assessed. In short, exobiology is already a scientific discipline in its own right.

Chapter XV. An Ideal World

> "There are three stages in the discovery of science.
> First, people deny that it's true,
> then deny that it is important;
> they finally give credit to the wrong person."
>
> Bill Bryson -American writer and publicity-

> "The brain, whether we like it or not, is a machine.
> Scientists have come to that conclusion,
> not because they're mechanistic party poopers,
> but because they have accumulated evidence
> than any aspect of consciousness
> can be linked to the brain."
>
> Steven Pinker - experimental psychologist and Canadian writer-

Western civilization has walked ever faster in its technological development, marking recognizable stages under the principle of exponential growth -not linear or continuous- collected by various laws. Perhaps the best known of these is the Price Act, always cited as an example, although there are others, which states that the volume of information in the world, and especially applied to digital information, doubles approximately every 18 months.

The first industrial revolution, developed in England during the year 1786. The invention of the steam engine by Wyatt, produced radical changes in production systems and on this principle, the first great wave of productive development worldwide was energized. The incorporation of mechanical traction tools, both hydraulic and steam, allowed the development of the mechanical and locomotive fabric, also driven by steam. More than that, it was the first time in human history that energy accumulation was achieved. Between 1870 and World War I, the second industrial revolution began in England, spreading through Western Europe, the US and Japan. Electricity, the electric light bulb, the radio transmitter and the internal combustion car were developed, constituting the beginning of significant changes that remain today.

The third revolution has been listed as the explosion of intelligent elements. It began in the last 30 years of the last century, introducing into daily life aviation, a new area of activity that would lead to the space age. There were decisive changes due to the discovery of atomic energy, cybernetics, personal computers and information technology.

At the end of the century, specifically in the 1990s, the general use of the Internet began. The fourth revolution has been described as the application of the Internet to the industry, with the management of products through the use of digitization, information and communication technology (ICT's) and smart devices. These have facilitated the connection of networks, communication with machines, allowing the supply of products and services to customers anywhere in the world. Although some original thinkers, believe that globalization began with the coverage of the global geographic map, starring homo sapiens, certainly the digitization of communications and information, has sealed the unstoppable trend of globalization and this hatching does not apply only to the area of business, but it is one of its most productive areas of action.

The Internet, like any decisive technological advancement, reflects a complex picture full of lights and shadows. Its widespread use has modified the space-time structure that governed the global panorama in previous years; immediacy, as a fundamental feature has pulverized the ancient geographical schemes, especially those related to the planning of times and geographical locations. A group of large beneficiaries is made up of large international financial investment groups as well as private actors. Buy-sell operations are carried out automatically on the world's five major exchanges. Occasionally, this immediacy makes it possible to obtain large surpluses or the evasion of tax or tax responsibilities. In Africa, small private investors through a simple mobile terminal, perform previously unfeasible trading.

Business relocation, initiated with internationalization of enterprises, promoted and subsidized by governments, considered the global market in the planning of its basic functions. Right now, this harmful and sometimes illegal mechanic is enhanced by the increased possibilities of global communication. Large corporations are characterized by the dispersion of their elementary functions. The production process is located in those countries where the cost of labor is lower, even small. This includes the perverse use of minors in productive tasks, under denigrating conditions constituting criminal offenses. Major brands, especially belonging to the sports field hide under their friendly advertising policy, a manufacturing process, located in Asian enclaves that use child labor.

With regard to commercial distribution activity, real-time communication with the various logistics platforms also at international level, allows to circumvent old barriers, such as the so-called problem of the last kilometer, by which access to points fate, posed a serious obstacle. This applies only to companies that produce and distribute physical products. Obviously, those corporations whose activity is defined by their digital seal, for example, search engines -Google- or social networks -Facebook -, their success is complete.

Finally, this novel global framework supported by digitization emerges, a well-known fact, through which transnational conglomerates carry out a large activity of transferring funds and assets, both among their different branches located in several countries scattered across the globe, as among the extensive network of instrumental companies -screen societies- that are usually formed or bought, intended exclusively for tax evasion. Financial engineering, based on these exchanges of turnover or assets, allows, on the one hand, to circumvent the various government controls aimed at taxing, for example, the Tobin rate in France, while taxing in those countries -for example Ireland- which demand a ridiculously low tax rate.

The recent rise of a virtual currency, the bitcoin has violated but did not reach the shudder of the international financial system. Based on the methodology developed for secure chains -blockchain-, it allows economic transactions to be carried out over the network. Operations cannot be tracked by banking or government institutions. Traceability -the string that identifies origin, intermediate stages and destination- of any product or service, in normal situations, is hidden here; impossible to trace. As is easy to imagine, any consolidated power group trembles at the possibility of losing control of their vital business areas; all the more so as, managing money, it shows how the world moves, especially today.

The emergence and acceptance of bitcoin is due to the main trait governing the behavior around paper money, called fiat money. The etymological root of the term refers directly to its Latin origin: faith. In other words, paper has no value in itself, it works by the generalized convention on its value as a unit of exchange. Since 1981, in which US President Richard Nixon signed the order to withdraw the gold value as collateral for the dollar, due to the steady growth of US foreign debt, the paper currency only presents as collateral, the signing of the President of the issuing Central Bank, in the case of Europe, of Mario Draghi. Prior to the date indicated, the issuing central bank was under a legal obligation to exchange paper currency by real currency or gold units to any citizen submitting such an application. Similarly, Facebook announced the launch of its own currency, the "Pound", an initiative that has temporarily snare. There is no need to make an unreasonable imaginative effort to understand this temporary suspension: The opposition of political and economic powers.

The potential basis of all digital business initiatives is anchored to the existence of a phenomenon that modern marketing has already called: captive customer. This is undoubtedly the case for leading digital corporations, specifically the GAFA group -Google, Apple, Facebook and Amazon-. Google is linked to a wide variety of technology initiatives; while Amazon now dominates 60% of the digital sales market. Facebook, although with greater commercial threats, due to the emergence of competition on social networks, maintains an impressive number of affiliates; Apple has positioned itself as an undisputed reference, in terms of representative lifestyle of upper and medium upper social class, especially due to the Iphone.

All these facts represent a range of initiatives perfectly explained by the neoliberal mentality, dominant in today's global economy. But the increasing digitization, makes it possible to appear activities of negative sign. They are mentioned here, perhaps the most representative ones. The emergence of cybercrime, represented by the admired yet criticized, figure of the computer hackers -it is a term without literal translation into Spanish, replaced by pirates-. Without having to go down in detail, the "hackers", acting either in a group -Anonymous, perhaps the most well-known case- or individually, have launched attacks on the central servers of all kinds of institutions and organizations, reaching as far as government institutions. The reasons given for this type of action lie in the simple rejection of the dominant political, economic and cultural system, until the demonstration of computer skills and, of course, economic crime. Faced with this state of affairs, in April 2018 the "Internet Forum of Industrial Security and Security Protocol" was made public. It established the most effective set of best security practices to be incorporated into all implementations of Industrial Organizational Intelligence (IO). It was not a matter of method or simply formal, the classification of the Internet as a democratic platform is broadly true, representing a decisive element of analysis in shaping global social structuring.

But its use for malicious or criminal purposes attacks frontally against values of presence essential in the oriented collective development, such as security and the desired feeling of calm coexistence, expressed by much of the Society. The progressive activity of cyberterrorism has plunged entire societies into chaos and disorder. Apart from organized terrorist groups, the figure of the isolated extremist or voluntary martyr has emerged. A growing number of young people, usually without training or a clear future, have adhered to such attitudes in large part, under jihadist propaganda from a huge number of websites and radicalized media. In large part, they belong to indigenous families in the countries in which they are part, in their second or third generation. The reasons, it is clear, that they do not obey religious influence; have to do with higher-spectrum causes, such as future expectations or the lifestyle expected by most of these young people.

"Dominating the Fourth Industrial Revolution," stated the guidance title of the World Economic Forum held in 2016, attended by some 2,500 people, including heads of states, entrepreneurs, academics, representatives of agencies outstanding members of civil society. The "XI Global Risk Report 2016", prepared for the Forum, developed a long series of threats considered to be the main significant challenges. A first group included climate change mitigation and adaptation, weapons of mass destruction control, the water crisis, unemployment and large-scale involuntary migration. A second chapter included commercial and social factors, such as rising energy prices, a lack of global governance, the generalization of corrupt practices, such as tax evasion and the existence of tax havens intended to provide shelter for this kind of widespread practice, the risk of past and future banking debacles, asset bubbles and, above all, businesses covered by idioms and scams. In the last label of this long list, was the great danger posed by cyberattacks that reach large industrial corporations and, above all, public institutions, including national governments.

Digitization, easily understood, consists of the process of converting communication material and analog information - for example a book - into a format, understandable for a computer working with digits -1 and 0 -binary code. It is therefore the transformation of continuous information to discrete or discontinuous format. In 2002 year of the generalization of the Internet, 99% of the world's information had an analog format; in 2008, that same percentage, corresponded to digital information.

The United Nations reported the number of Internet users, estimating it at 3.2 billion people -just under half of the world's population-. Today's age, known as the information age, inexorably leads to innovation in state-of-the-art technologies that drastically transform the range of human activities, from production, marketing, distribution and logistics and above all, the offer of new products and services, shaping a new type of business activity, friendly capitalism, based on two fundamental axes: The voluntary participation of users of platforms, websites and social networks, providing private information, both by the upload to the cloud of personal information and compliance, by awarding the typical "like" and authorization for the use of "cookies" - the user's metadata. All this data is too often used by digital companies. Auctions are often held, for the illegal sale of user references, intended for the use by other companies, offerings of any type of services. Personalized advertising about self-interest, as well as the awkward phone calls that any Internet user has received, are rooted here.

In parallel with the implementation of new trends such as three-dimensional 3D printing, robotization of production and drone distribution, resulting in a series of undesirable trends. The abandonment of rural areas towards large urban concentrations, the destruction of ecological reserves and above this, the acceleration of changes, which governments slow in generating effective policies will determine largely, the uncertain global outlook nearby. In the first 5 years following the current date, an estimated 7.1 million jobs in the 15 largest economies on the planet. In addition, 2 million hard-to-occupy jobs will be created, due to the demand for skills and skills, different from those traditionally requested. In different business sectors, the displacement of workers will be accentuated, due to the replacement by smart devices, in particular in health, energy and finance. The three main skills, considered in the 2020 Workplace Forum, are the competence in solving complex problems, critical thinking ability and creativity. It should be added, training in new technologies.

During the 230 years of industrial revolution, societies have undergone all kinds of conversions, with medicine being an example of its contribution to general well-being; anesthesia, x-rays, antibiotics, tranquilizers, electrocardiograms, DNA, vaccines or stem cells are some of the most relevant and decisive discoveries and applications. Despite general progress, in several countries the use of preventive and curative medicine is considered a privilege for people with financial resources or generous health insurance. For example, in Nicaragua private medical consultation and treatment for one month of any disease, i.e. diagnostic examinations, laboratory techniques and medicines, can reach an economic cost equivalent to three minimum wages.

There is widespread consensus in the overall appreciation of social progress; Despite the contributions of the four industrial revolutions, society and its rulers have neglected true human development, which should enhance a particular type of values such as peace, equity and social welfare. Since the first great human society, there have been many war conflicts, genocides and revolutions caused by ideological, religious or racial differences, with an embarrassing balance of millions of dead and maimed, as well as destroyed cities. The direct beneficiaries of this hostile whirlwind have been mainly the arms industry and transnational construction corporations. The presence in the real direction of public institutions, of shadow leaders, not democratically elected, translates into corrupt governments that allow organized crime, in exchange for economic compensation. Such activities have been so installed in the world economy that illegal businesses outperform the total legal business in economic volume. Thus, trafficking in human beings and trafficking, drug production and consumption, the sale and distribution of weapons, are some of the businesses that topseas the generation of global economic benefits.

Responsibility for this situation can be attributed in honor of the truth, to governments and public institutions but also inexcusably, to the lack of organization and demand for participation of civil society. Historically, great social achievements have been achieved after great episodes of struggle, sacrifice, and harsh opposition. The submission of legislative power to economic power, reserving a privileged position for the ubiquitous religious power, constitutes an invariable mode of social action, present since ancient times. The impunity of corrupters and corrupts creates a more stinging truth than corruption itself. Without entering into a thorough treatment of this point, which must be the subject of a monographic analysis, social iniquity, economic inequality and the disregard of individual dignity, have marked the path towards an already defined global situation.

The conversion of the global economy into a summage game of 0 means that the constant enrichment of the few in a globalized economy requires the forced existence of a large disadvantaged majority, not only measured in terms of poverty, but more exactly, of the misery present in the lives of the majority of the world's population. In particular, 1% of the global population has more wealth than the remaining 99%; the world's 62 richest individuals accumulate more resources than half of the general population (OXFAM International, 2016).

According to reports from the Organization for Economic Cooperation and Development (2015), the gap between rich and poor reached the peak in developed and emerging countries: 10% of the richest countries currently have incomes 9.6 times higher than 10% of the poorest; in 1980 the ratio was only 7.1 times. The globalized world employs 1.6 billion workers with stable jobs; 1.5 billion seasonal; and 115 million children working in dangerous and of course unacceptable conditions; in addition, 21 million victims of forced labor must be accounted for. The outrageous figures continue with 621 million who are not without work or study -in Spain called "nini generation". The International Labor Organization (ILO) believes that this critical situation has been caused by a slowdown in economies, inequalities and social conflicts. In his 2015 memorandum, he estimated 197.1 million unemployed people, 72 million of them under the age of 25. The World Bank's 2013 World Development Report reported the need for 600 million new jobs over the next 15 years.

Since the private sector provides 85% of jobs, public policies with a high degree of efficiency and equity should be designed and implemented jointly with governments, with solutions to the structural crisis that determines the current level of unemployment. But not only, action needs to be taken in this direction. Minorities with higher wealth rates should contribute in greater percentage to solving the problems affecting the weakest; after all, their fortunes have not been obtained in the lonely wilderness, but have required the collective in which they live and which provides them with the well-being and privileges to enjoy. A leading specialty in social progress comes from Biomedical research. Since the 10th century, discoveries, particularly in the area of vaccination, have saved millions of lives.

The first generation of this application, was produced in China with the inoculation of bovine smallpox, moved to Britain in the eighteenth century. The second generation of vaccines was developed in the 1880s by Louis Pasteur, aimed at eradicating avian cholera and anthrax. Since then, the use of vaccine terminology has been recognized. During the 19th century, the application of vaccination became widespread. The eradication of rabies, tetanus, diphtheria and plague was achieved during this period. The twentieth century has undoubtedly represented the time of the greatest discoveries. Whooping cough, tuberculosis, yellow fever, influenza, polio, measles, mumps, rubella, chickenpox, pneumonia, meningitis and hepatitis B and A are some and not all of the new contributions.

In the present century, research has not only not slowed down, but has faced more complex genesis diseases and some pandemics. The first vaccines for human papillomavirus have been obtained and pioneering contributions aimed at preventing widespread addictions, including heroin addiction and cocaine. It is also a question of confirming the total prevention of hepatitis C and influenza A.

The confluence of the different technologies allows great advances, both in the diagnosis, and above all, in the prevention and treatment of a wide range of diseases and dysfunctions. It has previously been noted (1) that computer chips had assumed some of the work previously reserved for human cells, especially specialist units in the immune system.

The current trend is focused on obtaining personalized attention, i.e. finding the best treatment or prediction -prognosis- for each patient who is at risk. This applies both to biomedicine through chip implants in different areas of the body, and to internal prostheses that experience significant advances in both cases.

Alamar, F. (2019). The Immortal Society: Towards Human Metamorphosis. Salamanca: Editorial Amarante.

The case of its use by so-called oversized or bionic men represents a separate section. In this sense, there are outstanding examples. Among them, the brain interface contribution of the tireless Elon Musk. This is not a purely progressive alternative; according to a specialized magazine, the company "Neuralink" owned by the entrepreneur, will report a business that will amount to a volume of 12 billion euros. Various research centers, including Stanford University and the University of California, have achieved the movement of a computer cursor, through a microchip implanted in the brain, although this goal had already been achieved in 2000. From this point on, communication between two brains has been achieved, with similar implants. Musk's method is really invasive, since it requires the implant of more than 3000 electrodes in the brain.

A use of nanoparticles has been developed by the National University of Rio Cuarto of Argentina, in this case, specifically aimed at the fight against cancer. These nanoagents recognize and only destroy diseased cells. It can be used as a pre-step to photo-dynamic therapy, which attacks only diseased cells and if tumors are inoperable. Another new method consists of the development of drugs, synthesized in the laboratory. An added technique uses microfluidic chips, as a liquid biopsy, that detect harmful cells, called exosomes.

Diagnostic and prevention protocols are combined with 3D printers for the design of physiological components such as a trachea cartilage ring using bio-tint as a basic building material. The team of Dr Adam Perriman of the School of Cellular and Molecular Medicine, under the University of Bristol in the United Kingdom, using stem cells, has achieved this leading achievement. Your team could get the impression of complex tissues, using the patient's own stem cells for surgical cartilage or bone implants, which in turn could be used in knee and hip operations.

Applications developed by telematic instrumentation, identical to those used in mobile terminals, are being used to save lives in Africa. Doctors Without Borders, he uses one of them to detect fever, the main symptom of malaria, as well as other conditions. Through this methodology, those cases that require an urgent transfer to the hospital are located. All this, in territories that also incorporate the added danger, of the presence of armed groups. This option has also served to examine 24.000 cases of malnutrition in children, of which 14.000 have required being transferred to the hospital. Likewise, with offline applications have been detected in Niger or Mali, children with acute malnutrition using a simple algorithm, which can be interpreted by volunteers without medical training.

In the same vein, Amref Health Africa has launched a new platform "mhealth or mobile health". Messages between medical experts and volunteers located in each village can combine written data and audios. Likewise, the car company Nissan, has developed an interface - dual device - brain and car. In the same vein, Facebook finds itself working on converting thoughts to text. It is implicit in the valuations and possible uses of AI - artificial intelligence - its application to social design. On the one hand, the predictive power of intelligent systems is indisputable, satisfying the ancient human desideratum of predicting the future and eradicating uncertainty. In the area of business, predicting the future has been a frequent activity in advanced organizations. Expert systems -former computerized programmes for the calculation of probabilities- in this regard represent the direct antecedent of AI. Essentially, they replicate an identical strategy; consulting for a sophisticated computer program, which contains a complex algorithm; thousands and thousands of lines of computer code.

Similarly, although without formal explicitity, some specific sectors of theoretical critics hope that the utopia of equity and total justice would be achievable through the use of an intelligent tool, which can value thousands of options, choosing the one that offers the highest chance of success. Would this be possible? The description of the company has already been operationalized by mathematical equations. The various constituent parameters of social structuring are included in the computer system in order to outline a particular social configuration and its numerical expression such as the behavior of the cohesion index or the spread of common behaviors such as racism or xenophobia. There are even free software for use for identical purposes with minimal knowledge in the handling of computer tools -Netlogo-. So the central question of increasing general well-being lies not in the technical possibilities, but rather in the will of the leadership group and on the other hand and as opposition, of the organized social community, that is, of the representativeness and strength of international opposition movements. In other words, another clear example of "parent´s home syndrome."

The discussion of the suitability of a given social structure has been revealed as a constant in sociological theory. Probably one of the most representative examples has been the clash between Niklas Luhman and J. Habermas, over social modeling, since 1971 in Bielefeld, Federal Germany. Both philosophers are committed to a holistic -global- view of social reality, thus distancing themselves from methodological individualism and the theory of rational choice, to which the latter gives rise. They are interested in analyzing the reality of the capitalist societies of the present day. Likewise, they recognize that such societies have reached such a degree of complexity that traditional sociological theories are unable to explain. For this reason, Habermas proposes his critical theory of society, while Luhman advocates for the general theory of systems. However, their disagreements are not only theoretical, but transcend social reality and materialize in the political sphere. Both positions are irreconcilable.

Habermas' formulation is part of the illustrated sociological tradition, that is, in a tradition that bets on the appeal of reason, in favor of human emancipation. It relies on the ability to achieve higher levels of humanization of the human species. Luhmann, on the other hand, advocates the abolition of the will and desires of citizens in order to obtain social order.

The great dilemma of the modern state in advanced societies is the pursuit of the loyalty of the masses, while avoiding their active participation in political activity. Morally undesirable, if politically achievable, goals. In this vein, the conclusions drawn from the progress made in the study of dynamic systems away from balance do not offer flattering prospects. The social structure is actually defined as a complex system, marked by deep relationships, a defined type of network. In this sense, it is subject, like the rest of the category of elements interacting with each other, to the emergence of power laws. All related systems, understood as the connection between a certain number of elements, whatever their nature, experience identical concentration phenomena. From the power grids to those of social type or the relationships of any real or virtual group, they are subject to the laws of power.

In any type of relational space, a minimum number of components receive most connections. An electrical network accumulates peripheral connections; on social platforms, influencers receive the highest number of visits; in the scientific literature, a small number of authors accumulate the largest number of citations. This phenomenon of concentration is called the Law of Pareto or also, "Law of 80/20" and simply illustrates the fact, that in any type of complex system a minority of agents or components, are responsible for the changes that have occurred in the system and also for its operation. Obviously, the percentages are not accurate, they simply serve as an illustration of the fact that a certain minority of the component elements are prominence.

In particular, in the economic subsystem, the continued repetition of prevailing practices on the world market has created a vicious cycle, so that the most favoured minority will continue to accumulate wealth to the detriment of the large and growing minority Impoverished. This fact has been called the Principle of St. Matthew, because of the text of the Evangelist which in chapter 13, verse 12, prays verbatim: "For he that hath shall be given more and shall abound; and the one who does not have, even what he has, will be taken away from him." Matthew (13,12). Likewise, in San Marcos -cap. 4, vers. 25- and in St. Luke- caps. 8 and 19, vers. 18 and 26-, the same idea is repeated. The global economic system continues to illustrate the maximum representation of a 0-sum game, as already said above, in which for a few to win, most must lose.

This scheme is qualitatively far removed from alternative formats, such as the one corresponding to the "everyone wins" option, typical of game theory such as the prisoner's dilemma. If any reader doubts the actual choice of applying other economic formulas in today's society, the models of the collaborative economy or the circular economy, are acquiring very interesting developments. The English current of the sciences of complexity, provides another relevant fact in the attempt to improve society. Complex systems between them, the social whole, are only susceptible to partial sorting -POSET partially ordered sets-. Among other reasons, because problems arise called in unspeakable Orthodox logic and in sociology of complexity, intractable situations. This means that although problems are clearly identified and theoretical solutions are found, the cost of their practical and real solution is un-assumed due to their magnitude. Today's examples of this principle are the international financial system, global warming, hunger and poverty or ocean pollution. However, having established the principle of total impossibility of social law, this does not imply the impossibility of implementing the most effective conditions, aimed at achieving maximum equity, a necessary condition for stability -homeostasis- to which every complex system tends congenitally. Language and recognition of each other's role, configure two of the elements necessary for this total panoramic framework.

Subtract, in order to delineate a perfect or ideal society, use computational simulation. This option is justified in that the analysis of complex problems is not capable of being concluded by means of equations that allow a single numerical solution. This is the case, for example, posed by climate change. In these situations, the finding by some theorists, on the demonic distance between reality and the virtual idea, makes sense. The will seems to be blurred, as the only possible bridge between the two distant banks.

It seems to be concluded, then, that the only way to attend perfected or ideal scenarios is to produce those same scenarios. This is exactly the main task of augmented reality or virtual reality. A notable test is observed in the film production, in which the impressive visual effects, allow the shootings using the simple interpretation of the actors, and sometimes it is not necessary, since the protagonists are also virtual simulations. Computerized projections of effects, landscapes and backgrounds have virtually completely replaced the costly task of "outdoor localization". This fact applies to both cinematography and its close relative, advertising communication. The level of perfection and originality achieved by the short communicative pills of a promotional or advertising nature, are in some although small cases, spectacular.

By using stereoscopic glasses and stereo phone headsets, artificial intelligence successfully undertakes with remarkable success, one of the activities that define its application: the accumulation and processing of a large amount of data, with the ultimate goal of recreating a wide variety of possible situations, all of which are idealized. In this way, in a near time, a user of these applications will find at your fingertips to spend a day with the lively recreation of a world-famous character, visit the always dreamed tourist destinations or enjoy an intimate dinner, with your actor or actress Preferred. The therapeutic potential of such techniques should be assessed.

A higher level of significance of these options, focuses on the recreation of second opportunities. Except in isolated cases, it is a general practice in individuals in particular, reached a level of age or some after a critical decision in their life, to question what would have happened, if the circumstances had defined a course of alternative action; in short, answer the constant question of "what would have happened if..." Artificial intelligence here applies the calculation of probabilities, a task that adapts perfectly to its main activity. By extracting significant data from the probability threshold existing in multi-determination systems such as life itself, relating to one's own abilities, frequent friendships, studies, family climate and interests, you can easily project an always desirable alternative; a panorama in which the subject only has to decide, which of them he would have wished to take.

The errors of an employment decision that has led to geographical displacement and therefore the abandonment of a known city, with all the elements that entails; the choice of a type of study and the consequent work dedication; a wrong marriage proposal or acceptance of a pregnancy or its contrast, and above all of them, the avoidance of the causes that lead to a serious accident or death itself. All this, at your fingertips. The happiness chosen, the desire achieved, the illusion on demand, in short, finally, the scope and attainment, albeit momentary, of life in a happy world.

XVI. Immediate Societies

> "A stupid faith in authority is the worst enemy of truth."
>
> Albert Einstein -Austrian physicist-

> "No one can escape anywhere."
>
> Milan Kundera -Czech novelist and essayist-

Technology and its development have strongly marked biological evolution as much as social progress. Together with the study of power, they constitute the two headings that bring together the greatest number of specialized contributions. The relationship between the two labels, configures a two-way path. Man has defined technology with the aim of achieving coverage of his various interests, while the products obtained through this advancement defined himself and the social groups of which he was a part. Technological progress represents a historical constant and has been used to catalog and define epochs, under successive labels, characterized by prevailing technological initiatives or products.

The consequences of the use and management of technoscience clearly transcend individual results as well as the mere social plane extending its influence to a global and therefore civilisational framework. It is not surprising, then, that the disciplinary debate in the social sciences today orbits the kind of society that the techno-scientific application will generate, or at least, about its substantial effects.

Probably for some time, articles, communications and books on the denomination of reflexive society have proliferated, reaching an asymptotic level, that is, maximum. The etymological root of the term reflexivity comes from the Anglo-Saxon world. "Reflexivity" refers directly to the act of thinking, at the moment of occurrence of events, continuously, dynamically and subjectively. The reflexivity in its most accepted concept, consists of a turn accompanied by internal reflection, in the direction towards the very subject of the action, similar to the process of observing the image itself in the mirror.

Different theoretical and critical currents have emerged, with regard to the implications of the adoption and implementation of the various technological specialties widespread in the social sphere. Some of them take extreme character in their basic assumptions, such as the case of Bruno Latour and his followers, laying the foundations for a strong research program and the requirement for recognition and legislation on machines and artificial products, commonly used in contemporary human action. Ulrich Beck, a deceased German sociologist, developed his career at the University of Munich and the London School of Economics. Among his many subjects of study, he coined the term of the risk society, to refer to the social, economic and industrial threats that may occur in situations where social control and protection institutions cannot develop the ability to take on and direct current technological changes in a globalized world.

According to your analysis, the type of damage caused and its significance can be considered irreversible. It is a sociologically brilliant conclusion, the consideration of organized movements as the new legitimization of a society characterized by disenchantment and the perception of existing inequality, which enhances the behaviors of individualization, a form of accession and identity, in a society perceived as threatening in many respects. In this reflective component, society becomes a danger to itself. In risky situations, one's self-awareness determines being. In this sense, the opposition with the greatest volume of components and significance, is starred by the anti-globalization movement.

This formation is actually composed of a set of activist currents who converged on a common position at the end of the last century. With regard to its name, many proponents of his action prefer the naming of an alter-world movement, in order to avoid the definition as opposed to the neoliberal economic movement. With a marked international character already in its beginnings, it was born in the World International Forum, held in Brazil in 2001, with a strong representation of opposition platforms, operating in Latin America. Under the generic motto of "another world is possible," it actually advocates a shift in the current values that energize the globalizing process, protecting the economic interests of large commercial and financial corporations and in general, of groups of that protect these same groups at the heart of international capitalism.

They hold the defining axes of current globalization, the emergence of the labor precarious, as well as the maintenance of an unjust and unsustainable development model. Moreover, the primacy of the influence of international capitalist groups seriously undermines the autonomy of world states, with the consequent deterioration of social democracy.

The declared enemies of the opposition movement, in addition to multinational corporations, have been identified as the direct cause of the existence of international debt, the main reason for the economic slavery of the countries euphemistically referred to as developing countries, i.e. poor countries. The World Bank (WB) and the International Monetary Fund (IMF) through misleading "development loans" have permanently chained underdeveloped countries to the obligation of an almost eternal debt, the demands of which they cannot bear; they do not even even have the resources to settle the interest on the new debts, requested for the payment of those previously incurred.

Various types of organizations, with different objectives even if confluencing, currently make up the primary cutting of the most recognized international opposition movement. Among them are in their own right, the anti-capitalist movement, which encompasses different political options, such as communist, anarchist and autonomist groups and parties. Closely linked to this option, emerges the working-class alternative made up of trade unions, labor movements and student organizations. Another stream of greater weight and tradition is formed by environmental platforms and sustainable development, supported ideologically by the numerous platforms of pacifist and anti-militaristic, anti-racist and indigent orientation.

The proposals made by the platform are therefore, diverse but consequential, with the general objectives of the movement. The forgiveness of international debt includes an already historic claim. Alongside it, it calls for the imposition of the Tobin rate, which is operational in some of the countries of the European Union that is levied by tax, economic transactions carried out on home national territory by large commercial and financial corporations. The proposals for change with greater significance are directed directly, to the necessary transformation of mentality and social vision, as an inescapable path to a society with greater doses of equity and social justice. The replacement of the generalized indicator of current economic growth, represented by Gross Domestic Growth (GDP), is requested by other indicators, with greater content, with respect to the relevant areas present in the social reality also requiring an increased social and environmental sensitivity. In this line is the proposal to energize the Sustainable Economic Welfare Index or the Human Development Index. At the very least, it is intended to be used in conjunction with current practice and is insistently calling for greater political attention to the importance for the current generation but above all, for future generations of the ecological footprint.

The package of proposals arising from the monitoring of the measures contained in the sustainable development objectives would, in a clear overall benefit, among them and especially those affecting food dignity -avoiding the unsustainable death, hunger and current poverty ratio, as well as the right to free movement of people that would free. Probably, the stream of alternative thought with greater intellectual repercussions has starred the sociologist, philosopher and essayist of Jewish origin, Zygmunt Baumann. It is formulated under the proposal of the concept of liquid modernity. His objects of study have included decisive social axes such as socialism, class struggle, poverty and globalization. In the late 1980s, there was a decant to a dystopian -undesirable utopia- view of the social collective of that time. Liquid society refers to the concepts of variation, fluidity and adaptation to the continuous alternatives that occur in the coexistence habitation field. It is the use of a metaphor, since the author recognizes that such alterations are profound and irreversible in nature.

This vision is in frank opposition to the society of the past, which could be characterized by solidity in its practices and routines, as well as by its contents and values. Like Baumann, he analyses reflexive society, concluding the need for a deep overturn. In our view, its contribution of greater social significance lies in the consideration that the greatest problem facing Western society concerns the coexistence and appreciation of the other, of the one who is valued as different from us. The main strategies used in the relationship with the alter-ego are the epic strategy consisting of its exclusion from the environment considered to be its own and proprietary. Taking on the different individual, but stripping him of the traits that characterize him as such, that is, destroying his otherness, constitutes the philological strategy. Both methods constitute a mode of invisivilization of the different, until the moment when it completely disappears from the own mental map.

Liquid society is characterized by the renunciation of individuals to security, in return for higher doses of freedom. The real problem with these two concepts -security and freedom- with the ability to define a social system on their own, is that they are at the same time incompatible but mutually dependent.

Freedom can hardly be attained, without the necessary perception of security, desired individually. In turn, life autonomy is the second right, to the highest degree in the hierarchy universal scale of values, preceded only in importance by the right to life. Historically, societies have tried to strike a balance between the two parameters. The idea of the welfare state reflects this attempt and its appearance is relatively recent, specifically it appeared in the second half of the last century. The manifestations included in this social vision are in the rise of universal health or universal suffrage. Franklin Delano Roosevelt, at the time, expressed the need to liberate society from fear caused by insecurity.

Today's social communities are characterized by being open nuclei, but with a lot of artificial burden and manifest fragility; receptive to information, migration or capital. Thus, that in a globalized world it is a relatively simple objective to impose a recognizable mode of imperialism; this attempt is personalized in the United States of America. But due to the congenital duality of complex phenomena, the pretence of cultural uniformity entails a greater degree of opposition and resistance. The emergence of terrorism, in all its expressions must be attributed in part to this attempt at forced uniformity. It is not known whether Newton in formulating his third general law, which expressed the appearance of a reaction of equal intensity but contrary to the pressure exerted on a body or system, thought at some point that it would have application on the civilizing plane 400 years after its formulation.

According to the author's vision, in the face of the existence of the vicious circle, characterized by institutionalized unprotection and unbearable insecurity, loneliness and helplessness, do not constitute acceptable situations. The construction or deconstruction of a true re-imagined pedagogy, supported by freedom and democracy, justice and equity, seem to be an alternative not to reproduce but to emancipate; not to control but to unleash, because "the most dire consequence of the global triumph of modernity is the acute crisis of the 'human waste' destruction industry.

The combination of the principles of dynamic systems, together with the geometry and mathematical analysis of social communities, has led to a theoretical current that came to make a great impact, starting with their diffusion in the early 1970s. The French mathematician René Thom, using differential calculus, in particular morphogenesis elaborated the theory of catastrophes. This model has its application in the analysis of social evolution, both systemic and mythical. In essence, the theory expresses the tendency of structurally constituted stable systems to manifest discontinuities and discrete changes -non-linear or continuous- often suddenly, which can lead to involuntary processes, among other possibilities. Fundamentally, it handles well-known assumptions from dynamic systems, such as reliance on initial conditions, the emergence of branches in increasing the complexity of a system, and the necessary phase transition.

The term catastrophe applies above all to instantaneous variations called singularities. At this critical time, structural collapses can occur, such as economic activity, the collapse of the financial system. Predictable changes are seen in the immediate future, the seeds of which are germinating at present. The incidence of global global warming, in which climate change represents its spearhead, has previously been mentioned. Unlike a list of issues that frame imminent changes, but will be progressively resolved such as cyberterrorism, blockchain, the influence of robotization at work, energy transformation or the necessary adaptation of training of future workers, some imminent problems are more complex to solve, due to the demand for this, of complex and more important measures. This is the case, of the reversal of the population pyramid, that is, the unstoppable propensity towards a global sexagenarian society.

The progression will increase, catalyzed by the progressive increase in average lifespan. Thus, the most well-known pyramid shape varies in shape, to a more similar image to a cube, in which older groups increase their volume. One of the most obvious risks of this picture lies in the pension financing model, up to the hour based on the direct contributions of active workers, earmarked for the subsidy of already retired segments. At this point, for example, in Spain the ratio is set in the 2.3 workers for 1 retiree; however, the trend leads to a ratio of 1x1, i.e. an active worker will be responsible for the maintenance of a retiree. Clearly, an insufficient equivalence.

In addition to this, the need for the care of the growing group of elders will require professionalized attention since until now, this help came from the immediate family environment. Despite their limitations on the effective plane, so-called social robots can perform another of the tasks of great social value in this particular field. Careful and complex programming could give them the possibility to interpret different gesture and bodily manifestations associated with the existence of different needs or physiological states, such as tiredness, pain, hunger or thirst, among others. Once the need is identified, the "company robot", like a lazarillo dog would target the different urban points that can meet the specific need detected, such as utilities, canteens or health care centers.

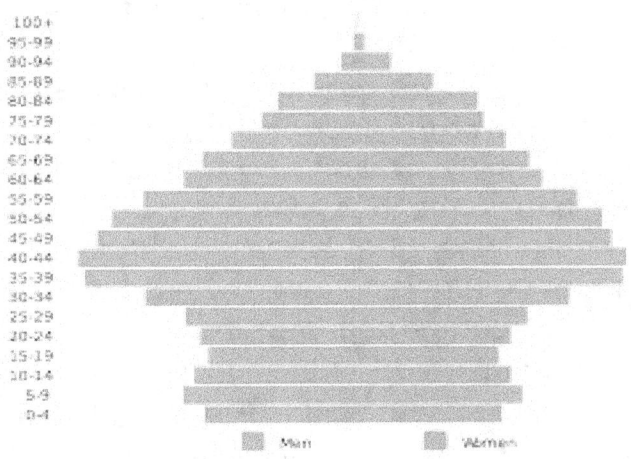

The population pyramid will vary its traditional hierarchical form in a continuous trend towards progressive aging where groups over 40 constitute the highest volume segment with the dangerous implications that this fact entails, for example, for the maintenance of the current pension system. Source: ca.wikipedia.org

The immediate future is outlined, through clear guidelines of profound impact. Undoubtedly, the persistence of social iniquity assumes the axis of greater future prominence. The concept of social equality probably originated in Aristotle and was defended mainly, by the religious current of Tomist origin -St. Thomas Aquinas-. Its evolution achieves with the French Revolution, its universal projection and consequently, its access to the political sphere. Subsequently, the fundamental rights of the individual, such as freedom of expression, legal treatment through a fair trial, the right to vote or religious freedom, obtain their final endorsement in the promulgation of the "Letter of Rights of the United States", signed in 1791 and assumed by the State after World War II, supported by the ratification of the "Universal Declaration of Human Rights" promulgated in 1948. Today, economic behaviour produces serious effects rightly classified as unfair and harmful to genuine social development as well as to the environment.

The second defining axis of today's civilizing system lies in the generalization of the use of the Internet, as part of a global process of digitization. This conditioning influence completely shapes the delimiting lines of specific social groups. In this way, the so-called Generation Z or I generation has emerged, formed by individuals born at the end of the last century and who today do not exceed 25 years. This group does not conceive of life without permanent access to the network through some kind of digital terminal, usually the mobile phone or tablet. They can be perfectly defined as digital natives.

The values prevailing in this generation, are logically transferred to your work activity. Passionate about change and stimulus, they constitute the largest core of entrepreneurs and freelancers who collaborate mainly with companies with an advanced degree of technology. They represent the main drivers of start-ups -young companies with innovative ideas-business-. In the workplace, they share space the so-called "5 generations of workers". The generation immediately preceding, the so-called millennial's formed by individuals born from the 80s and therefore do not exceed 40 years. Sociable and flexible, with access to digital skills they rely on their work as a way of life, which facilitates their vital independence and of course economic.

With a similar profile, the segment under 45 years of age is defined as generation X. Adaptive, free and informal, they maintain a recognizable code of conduct, especially in the area of work in which they demand transparency and coherence, without renouncing the questioning of the authority, in the event that it clashes with its principles. In the upper age stretch, specifically, between the ages of 55 and 64, are the descendants of the baby-boom. Normally, they have achieved professional success and the added privilege of economic well-being. They have shown doses of ambition and risk but with the demand for immediate economic and social reward. Finally, there is the oldest age group, formed by those who continue their work voluntarily, either by habit or for reasons of profession, usually autonomous or liberal. They are accompanied by a higher dose of acceptance and ability to defer rewards. With conservative ideology, they defend the current state of relevant issues and accept the rigorous discipline of authority.

Yet the seemingly unstoppable trend lies in the advancement of technological confluence, which through new sets of scientific applications seem to lead to the immediate society, to a real paradigm shift of civilizing breadth and depth. The acronym NBIC reflects the confluencing development of four of the disciplines whose achievements are exposed throughout this writing. This acronym groups the initials of its applied references: atom, gene, bit and neuron. This gives its own space to four emerging disciplines: Nanotechnology, Biotechnology, Computing and Artificial Intelligence and Nanotechnology.

Nanotechnology enables the use of engineering on an atomic and molecular scale. Linked inseparably to the research of new materials and energy superconductors, it has enabled the manufacture of new compounds, valued by some critics as "on-demand materials", for the first time in history. These new components have specific properties that are controllable and used for very specific purposes, theoretically always aimed at improving the quality of life and social progress in general. The possibility of molecular diagnosis that makes it possible to identify diseases in their early stages has already been highlighted, at which point it is possible to eliminate them definitively in a relatively simple way. By using different types of nanofibers, it is accessible to achieve the growth of artificial tissues and organs, from cells of the individual or patient. Particularly attractive, its application arises to the process of technological transfer, aimed at environmental safeguarding. Advances in cell or solar cell design, increased efficiency batteries or wireless electricity transmission point to some of the current working directions in this area. Different laboratories, researching the manufacture of paints that deposited in windows, acquire the role of small photoelectric plants, capable of trapping solar radiation for later conversion into electrical energy.

The human genome has once been described as the "book of life." The great compendium of coded instructions for each kind of planetary living being. From its complete interpretation, the connectomics -the attempt to connect nerve cells -as well as the optogenetics- the technique capable of illuminating the desired connections -the bionic- the combination of cybernetic components to the body- has been developed and the brainnet project, the direct connection between human brain and the internet.

The second emerging acronym, represented by the acronym BANG -Bit, Atom, Neuron, Gen-, represents an alternative approach containing essentially identical components but represents the ambition for greater achievement, as it aims to obtain all technologies used, in a nanotechnological format; on a nanoscale that handles magnitudes a million times less than a millimeter.

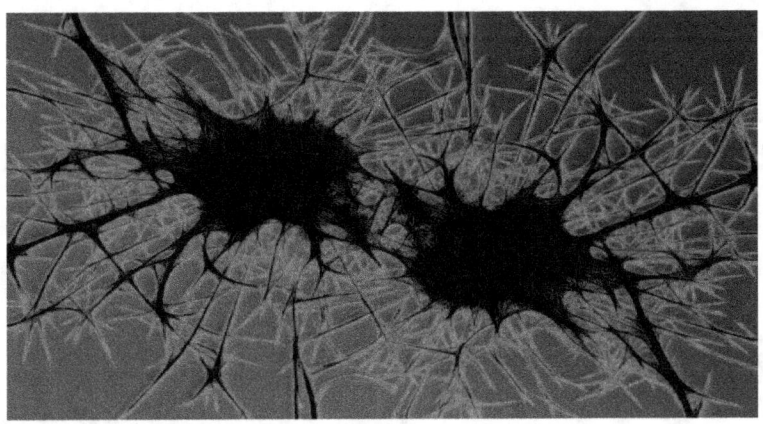

Nanorobotic will achieve the desire of biomedicine focused on the prevention of degenerative diseases as well as diseases caused by bacteria or viruses. A nanofilament a thousand time thinner than a human hair will act as a biological "police" detecting any external body considered potentially harmful and proceeding to its elimination.
Source:Pixabay.com

Various U.S. political initiatives, along with a group of head industrial corporations, participate in the "Manhattan and Apollo" projects. The latter, with the aim of achieving the nanoscale in the technological confluence. According to this group, if this goal is achieved, it would lead to the greatest industrial revolution of all time, since it would imply a "rebirth" of society as it is now known, evidently led by the United States of America. The National Science Foundation of this country (NSF) coordinates its efforts with the Department of Commerce and with the Subcommittee on the Integration of Engineering, Technology and Nanoscales Sciences (NSET). The implications of this alternative, according to a report by the North American Government, apply to atomic technology, as well as to food and agriculture, on which the latter depends. The Country's Department of Agriculture estimates a budget for the nanotechnology chapter, an increase of more than 900 %, compared to the average of previous years, logically encrypted at several tens of millions of dollars.

Chapter XVII. In the Kingdom of Heaven

"Nature never seeks intelligence until habit and instinct are useless. There is no intelligence where there is no need for change."

H. G. Wells - British historian and philosopher -

-"It's harder to break a prejudice than an atom."

Albert Einstein -Austrian physicist-

The impact of science on the whole of global citizenship has reproduced the opposite social groups, identified by the classical sociological typology. In this way, existing positions that are identical are clearly shaped, in all matters of great depth and general depth.

At the point of defending the choices and beliefs that express denial of the events or phenomena that are analyzed or dealt with, there are the denies. This group is characterized by the explicit rejection of any scientific progress, which depositions its conservative beliefs. Thus, climate change, trans-sexuality, UFO sightings or technological applications in humans, arouse a resounding denial. In the case of technoscience, they are called bio-conservatives.

On the opposite side are progressives, who in the case of technological innovations, add to their name, the specific bio prefix, that is, bio-progressives. In the face of bio-conservatives, those who advocate technological improvement driven mainly by the transhumanist movement are located, rising above the maxims of unlimited improvement and a frontal challenge to death and aging. This postulate could achieve its operational implementation, by virtue of the impressive possibilities offered by bio-technologies. The control of aging and rejuvenation, becomes for this broad social segment a moral duty that must be assumed by humanity and recognized in the context of fundamental rights, as noted for example, José Luis Cordeiro and David Wood . Among the authors and specialists who can be listed as bio-progressives are the philosopher Peter Sloterdijk, Ray Kurtzweil or Nick Bostrom.

This social division has occurred for at least 300 years, at which point Darwin proposed his theory on the origin of species. The idea that man has evolved from a common trunk, shared with the chimpanzee, created a majority movement of collective opposition, called creationists. The basic belief defended by this group is supported as can be easily anticipated, in the conviction that man and by extension, the world as we know it, results from a divine creation. Hence, any changes or progress are automatically rejected by the reactionary group.

On the opposite side are the defenders of the universal theory of panspermia. The assumption on which this assumption rests, focuses on advocating that the development of life is a general process, applicable to any place in the universe, that complies with certain rules. If life has developed in successive phases on Earth, from essential components produced by stellar explosions, this process will be repeated indefinitely, at the time under the principle of equal application of the physical laws in the whole universe, wherever the climatic conditions conducive to its occurrence arise.

The functional equation of life includes as a basic premise the decisive influence of the immediate physical environment -environment-, as an essential catalyst of evolution. Human beings, like the rest of the universe, would have their origin in stardust. This is a beautiful but challenging front conclusion of the assumption of the divine omnipotence, creator of all the conditions of the cosmos, by virtue of its full omniscience. In other words, infinite divine intelligence would have been responsible, depending on its full knowledge, for the possibilities of nature, for the decision to produce the so-called intelligent principle. The human being as the ultimate exponent of universal perfection, by virtue of divine grace.

Conservative groups set out principles that determine their negativist stance, totally in line with the ideology they profess. Thus theological objection is one of them and its main argument, focuses on the consideration of thought as a function of the immortal soul of man. God has granted a soul to every man and woman, but to no animal or machine. Therefore, none of them have the intellectual capacity. This argument represents a clear case of conceptual double fallacy. First, an unproven fact -the existence of the soul granted by God- is claimed, and then, to conclude an impossibility, based on a false or unproven premise. On a similar line, the so-called ostrich objection is placed. The argumentative process is clearly aspirational, but in no way objective. In particular, the statement similar to the fear arguments of political discourse states that "the consequences of machines thinking would be horrible. We believe and hopefully that will never be possible."

One more step in this same line of argument, it has been established by Anthony Levandowski, engineer who promoted the development of self-driving cars for Google and Uber. It proposes the creation of a new religion to "develop and promote the realization of a deity based on artificial intelligence and that through understanding and worship of that Deity, contribute to the improvement of society". This author strongly expresses the belief that God will emerge, specifically in the year 2042 and write a new Bible.

Far from hyperboles and conceptual biases, it is true to combine a number of elements in the appearance of life on earth, which may lead to the erroneous conclusion but not without argument, that this phenomenon must have been the product of a chance directed or an act based intentionally, i.e. finalist or teleological. For this reason, the discovery of life on any other habitable planet would end sharply with the supposed creationist. The main argument put forward by the creationists, focuses on the Earth's own characteristics, observed in relation to the general distribution of the universe. Its location at a point far away from the center of the galaxy, in turn, placed at an almost marginal point of the galactic center, has been argued as a weighty argument, in the finalist reasoning. I mean, why here? Second, the different times of the planet's shaping ended in climate stabilization, with the stable atmospheric conditions, necessary for the development of life.

Likewise, the existence of water, one of the key elements for the emergence of living organisms, is seen as a result of an intelligent decision. Water has a number of properties that make it one of the strangest elements of the entire known universe, along with dark matter. It is a fluid that has a very accentuated electrical conduction capacity, acting as an unmatched catalyst in the energy reception coming from the Sun. Likewise, the degree of salinity and acidity offer very wide ranges but at the same time, with a tendency to stabilize at medium levels. Its ability to maintain a temperature range, relatively independent of the outside environment, - except in extreme conditions - provided the right breeding ground for the development of life, while the planet underwent at least five stages of change, up to atmosphere formation and climate stabilization.

The unique properties of water, analyzed by means of x-rays and subject to another series of experimental tests, show unique results and characteristics, both individually and from a relativistic perspective, i.e. compared to the rest of the known liquids.

Water contains two particles of hydrogen (H2) and one oxygen (O), with different functions and which give it particularities, sometimes considered as anomalies. This particular liquid, presented a compact cube-shaped structure, in which hydrogen maintains strong internal coherence, while oxygen atoms tend to fuse with different adjacent molecules. Internal cohesion confers on it a condition of ordering, which becomes at a high surface tension, so that if two currents at different temperatures cross in the ocean -as is often the case- do not mix with each other; this facilitates the beneficial effects of global and cyclical phenomena such as the Current's Child.

Water behaves differently from the 70 most well-known liquids. Its unique composition allows it to be the only liquid that can exist in a stable way, in the three different states: ice, liquid and gas. Indeed, the different pressure and temperature conditions cause surprising changes in its molecular structure. The high temperature produces the decrease in the resulting ice density; adopting the anti-other liquid trend, frozen water decreases its density, allowing ice to float. This anomaly allows fish and other forms of aquatic life to survive in the waters beneath a frozen layer of ice, protective during cold periods. At the opposite end, it also shows a high caloric capacity; Having the dual ability to absorb and release a large amount of heat, stabilizing its average temperature level, ideal property for higher living beings, to maintain a stable body temperature.

The most surprising results on this common element in the cosmos were obtained as a result of research by physicists Anders Nilsson and Lars Petterson of Stockholm University and published in the journal Nature Communications. The research consisted, in the analysis of the results found by dozens of articles published in recent years, relating to their molecular structure, using cutting-edge research techniques and instruments and computerized simulations.

The main conclusion drawn is the proposal of an image of the liquid that presents a heterogeneous structure and even more surprising, a fluctuating organization that frequently changes state. In other words, water has a dual nature. In addition to this, there are numerous regions or spaces differentiated between these two boundary structures. The different extreme zones, with different ordinances occur under pressure limit conditions (P) and temperature (T). A highly neat, low-density configuration with strong hydrogen bonds and a high-density structure, slightly crushed with distorted hydrogen bonds. Thus, the image of water presents fluctuations, giving rise in its graphic presentation to a phase diagram. The intermediate point between the temperature and pressure variables shows a direct correspondence, with the identical conditions in which life develops. In this critical area, the branching phenomenon, typical of complex systems, is presented.

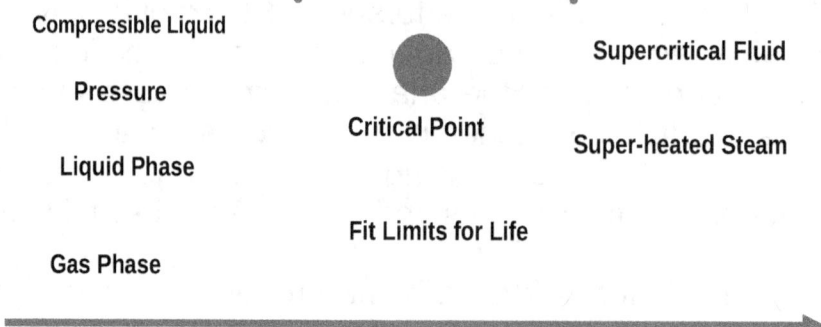

Changing pressure and temperature conditions show the malleable capacity of water, the unique liquid in the universe that can adopt those of steam, liquid, ice or gas, due to its constant fluctuations, so that can be interpreted as a liquid properties two liquids kines by a relationship the same type; provide suitable limits for the existence of organic life manifesting acritical point from wich the tipical bifurcation of complex systems arises. Source: Own elaboration.

The liquid has other relevant characteristics. It is a powerful solvent element, capable of acting effectively with a wide variety of substances, such as alcohols, acids, salts or alkalies. The adherent power is another of its main characteristics, so that it joins with great force the crystals, also at the macro level; two water-lubricated and simply bonded metal sheets are difficult to separable. Its mutation capacity is illustrated by the unusual fact that hot water can move into a frozen state, with greater speed than cold water. All this can lead to the attribution of metaphysical properties to the liquid element.

4 billion years ago, 20 major protoplanets were orbiting the solar system in formation. One of them, called Teya about the size of Mars, collided with Earth and then disappeared completely. The enormous magnitude of the impact, caused an imbalance compensated different reactions such as the inclination of its axis, the elongation of its shape and the increase in the speed of rotation. The Moon originated as a result of the collision, as demonstrated by the analysis of the composition of the lunar rocks. Currently, the distance to the satellite -just over 384,000 kms.- seems the exact location, to contribute to the balance of its mother planet. Indeed, the star exerts a significant influence on the dynamics of the tides and the maintenance of the general terrestrial dynamics. The Moon has exerted a mythical influence through the times, while becoming one of the arguments defended by creationists in the construction of their direct hypothesis.

Finally, the last argument about creative theology, sets his gaze on the size and position of Jupiter. The gas giant has a volume and occupies a position among the outer planets of the solar system, which seems to protect Earth from the frequent bombardment -not of all, of course- of meteorites that would otherwise hit the planet.

The arguments alluded to by the proponents of the universe's creative process have systematized an alternative doctrine, called anthropic principle. In its strong form, this assumption defends that the ultimate end of the divine creative process lies in the formation of man as the supreme being of creation, since it has been created in the image and likeness of God. The weak version proposes that the conditions on the planet were created so that the emergence of a creature capable of wondering about its meaning in the universe was possible. With regard to the latter formulation, a counter-argument emerges, but equally applicable to the whole doctrine of creation. It refers, to the supposed infinite goodness of the creator, in relation to the injustice and inequality that exists in the world. The official religious doctrine has found in the concept of free will, the perfect justification. Thus, God was the creator of the world and all that it contains, but subsequently allowed, to the magnificent act of creation, for the human being, to have freedom of choice with regard to his actions; and not only that, but in this way he could choose to follow the religious path of his own free will.

As it is a matter of such historical and social relevance, the decision between the generating choices of life -panspermia and divine creation- is resolved in a personal way. Science handles probability criteria, about these two options. One of the most respected principles in determining scientific truthfulness and validity, is called the principle of parsimony or alternatively, Ockham's knife. The rule explicites that, faced with equal probabilities, the simplest option offers greater options to be true. But as Franz Wefel rightly put it, "no explanation is necessary for the one who believes, for those who do not believes, all the explanations are left over."

Chapter XVIII. The New Apocalypse

> "... cynical recognition of a
> unjust global situation
> doesn't point to a deficit of knowing
> but a corruption of love.
> Those who might know best
> they don't want to understand"
>
> Jorgen Habermas - German philosopher and sociologist -

> "¿Has science promised happiness?
> I don't think so. He has promised the truth and
> the question is whether with the truth
> will one day achieve happiness."
>
> Emile Zola -French naturalist writer-

The triumph of rationalism, by which man placed himself as the ultimate central expression of creation, as opposed to the contrasting vision of the religious belief prevailing in the seventeenth century, led to a disastrous consequence as he formulated a contrived separation between man and nature; the very nature that conferred upon him existence.

The proposal to distinguish the phases of human development, by virtue of the possible stages of technological evolution, is conceptually unacceptable, for a variety of reasons. First, there are no differences of sufficient entity between the different industrial revolutions, to consistently delimit the various proposed episodes; between a typical telar of the first, an electric light bulb belonging to the second, a plane characteristic of the third and a smartphone, distinctive of the fourth. There are no defining elements of epochs that intrinsically affect the definition of humanity, enabling the transhumanist proposal.

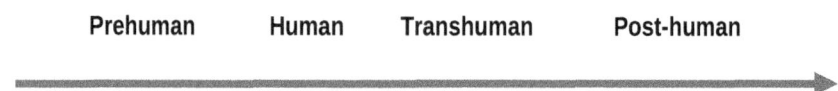

The four-phase distinction of human evolution incurs different conceptual and methodological biases. The rules of formal logic prevent the inclusion in a definition of what is intended to be defined and, on the other hand, the qualitative differences between the different significant products of the four industrial experienced by the western civilization do not offer sufficient degree differences to define a distinctive and less, unique state. Source: Own elaboration.

The human being does not pretend to be like God, as has exposed the well-known physicist and spreader Michio Kaku. Rather, it is the congenital inability to renounce this frenzied, irrational and deeply entrenched need to imagine, dream, know and explore; to know what is unexplored. Possibly, a feeling identical to that that led the most remarkable physicists in history to evoke God, at the time of discovering the fundamental laws that govern matter, as is the case of Einstein, Schordinger, Faraday and many others.

The distance between future reality and fiction, narrows the distance between it, but that doesn't mean it doesn't continue to exist. The ideal of the ancient alchemists, has been collected by the longed-for and admired sci-fi author, Isaac Asimov. In one of his short stories, titled "The Last Question", an intelligent supercomputer, learning for himself, becomes a certain moment, the last survivor in the universe. It thus expresses the final paragraph: "... But there was no man to facilitate the answer to the last question. It didn't matter. The answer, by demonstration, would solve that problem. During another interval without time limitation, AC thought of the best way to do it. Carefully, he organized the program. AC's consciousness covered everything that had once been a Universe and pondered what was then Chaos. Step by step, you had to do it, and AC said: MAKE THE LIGHT! And the light was made..." Fiction clearly, but unmissable dream for human beings.

So much so, that at a certain point in scientific development, advocates of the general method of science, probably tired of perpetual dispute with different religious doctrines, openly proclaimed that science was only applicable situations and areas where objective and replicable data existed. A clear example came from the continued reference, towards the reason for the big-bang; and inevitably towards its origin, in other words, the questioning of "what" existed before the beginning; ¿Improper madness? ¿Intellectual excess? Absolutely not. It can be admitted, although it is a stark limit, that the real question of science, is concrete in how and not in why, which corresponds to philosophy in its own right. But the love of knowledge and the very constitution of the brain, allied with a virtually inexhaustible spectrum of factors such as the instinctive fear of death, have come the long way up to a specific moment: technological singularity.

For this reason, the ancient ideals are pursued with the conviction of attaining them, or at the very least, do not cease their persecution. We therefore deny the affirmation of living a trans-humanism, leading to a new and later stage called post-humanism. On the one hand, for the labeling of history generates the desired feeling, inherent in conservative thinking, of continuation of an absurd unlimited progress. The division into decisive phases and the use of new nominalisms, such as extropianism -the belief of substantial change in human nature by external means- contributes to the consolidation of the conviction to follow the path right, in proper progress. Likewise, all the suffixes "trans" and "post", seem to obey the general desire to close or end supposed evolutionary stages, which have not been positively valued; is the case of the current era, called "postmodernity," in an attempt to overcome the mistakes made by the application of the meta-belief represented by the single thought.

Today's human being is no stronger or smarter than the first sapiens, rather on the contrary. Without going any further, the Neanderthal man enjoyed greater physical strength as equally, a larger brain volume than modern man. The cultural revolution replaced, from the very moment the first great civilizations were formed, the biological evolution, which seems wrongly, to have stopped. The acceptance of the four stages of humanity concreted in prehuman, human, transhuman and posthuman, represents a logical artifice, a fallacious construct and at the same time, a sociological presbyopia; it is fundamentally due to a historical bias. Different periods can be differentiated, as has been done, within evolution, but not include the defining concept of human nature, humanity as a stage in itself. It is a formal logical fallacy, as well as conceptual; what is intended to be defined cannot be included in the same definition.

On the one hand, there is no greater difference between a typical human and a cyborg, i.e. a human endowed with an artificial arm or with a chip implanted in his brain, than between a stone axe and a quantum computer. On the other hand, progress, although with a multitude of negative points, accelerates by becoming more human than ever. Yet the undisputed scientific progress of this century has been covered by an artificial mantle, which has enabled the replacement of the religious myth with technomyth, the myth promised by technoscience; some of their representatives, have drawn perfectly, a complete mystique of necessity; they have played with the ancestral human hopes of transcendence and channeled their legitimate irrationality, justifying it through the scientific method.

Technoscience and its applications have become the business of the future, perhaps definitive. The marketing strategy has not focused on the creation of new needs - acquired needs -, not even the consolidation of the old needs already created, but rather, it has consisted in the launch of a depth load, to the sphere irrational instinct; the ancient animal need for survival, finds a golden bridge, rationally and scientifically explained, towards perpetual survival. They have masterfully united, instinct and reason, science and religion. Energizing the basic pulses of individuals is always a safe bet on success; Donald Trump's campaign director and his advisers, they know that well.

Two major impacts, they emerge around the application of technological advances, in particular, the generalization of the use of Artificial Intelligence. The first is based merely on the possibility of achieving an absurd and random combination; in the specific ignorance of scientific methodology, coupled with the persistence of ancient myths, not to mention the perverse use of helplessness syndrome learned, equally persistent today, encouraged by both governments and by commercial corporations. The built or imaginary enemy, has been used for the proclamation of a permanent state of emergency, permissive with the violation of the most basic social rights, as well as used to facilitate the commercialization of all kinds of false remedies, from tranquilizers to "coaching" -training techniques to achieve success-.

The current myths come basically from fiction literature and religious tradition, together with Greco-Roman mythology, that is, its emergence, it has nothing to do with the emergence of robotics, nor artificial intelligence. In Jewish tradition, there is the legend of the "Golem", a being created from clay to which medieval rabbis lived life as a defense against anti-Semitic attacks. However, the recurring myth is found in the resurrection, expressed in many different ways, from the Greco-Roman "Ave Phoenix", "Prometheus" or its modern version, "Frankenstein". In short, the persistence of the myth already from homo sapiens; man is able to imagine something and his opposite, from the moment he acquired the enormous potential conferred by symbolic thought.

The possibilities of the future must necessarily be differentiated from the pure concept of probability, just as fiction must be distinguished from reality. Chance is an invariant part of probability, but the future does not necessarily imply improvement, it simply ensures evolution, that is, it enhances the role of continuous change, as an inherent rule of life, as already discovered by the Greek Heráclito. The second line of unreal promises is about the fiction of achieving eternal life, this time, riding a cybernetic construct. Again, the confrontation is defiant to the only law of universal application that states that every beginning, entails in itself an end, -yes, depending on different time scales-. These kinds of assertions not only constitute a false promise, but also and above all, a physical impossibility. But the harshest criticism is likely to turn to the cruel deception of irrational hopes, with the priority objective of obtaining clearly fraudulent and immoral economic income. In all these surreal panoramas, the metaphysical substrate of the dualistic vision of reality emerges, from its platonic systematization, in the "allegory of the cavern", one of the most representative works of philosophy within the work "La Republic."

Constructivism contains radical theorization about the imposition of culture or the ideal world, on objective reality, that is, on matter. It should be added that the definition of human beings as a "lack off" can be perfectly complemented by its status as a contradictory subject. Many of those who self-define themselves as agnostics fervently wish that their loved ones will be somewhere permanent, when rationality indicates that the only place where it can endure is in memory. Nietsche and Dostoyevsky expressed the well-known sentence "God is dead", to which the Russian writer ascribed "sin does not exist"; a literary form of expression of the denial of the myth. Nietsche concluded that "appearance, illusion, is a necessary budget for art as well as for life."

The application of science to human progress has led, as has previously been expressed, to obvious progress in the preservation of life, to the increase in health and the general well-being of society. This is not a numerical issue, even if a global spectrum of application is needed, including more disadvantaged peoples and groups, without access to essential goods and minimum services. As he well expressed in the film "The Schlinder List," "with a life that would have been saved, that would have been enough."

Metaphysical promises, scientific methods and therefore limited, the archaic confusion between myth and reality, the combination of desire and possibility, the new alchemists remain enclosed by the mystical halo. Roy Ascott argues that "the foundations are laid for one, overcoming matter, redefining categories such as reality, appearance and the real, is imminent. The molecular structures of our world are deconfigured, redesigning the atomic base of reality"; a phenomenon it calls "Nature II or its beta version". While a part of the assertions, the requirement of formal logic, is certain, limits the possibility of explanation of the whole by the party. Again, a mixture between acts of partial applicability and a literary metaphor.

The second promise of the new prophets of the future has focused the future possibility of a virtually eternal life, by incorporating a human and therefore perishable brain into a physical structure, on the other hand, in no way indestructible. A plant life, without the necessary stimuli for its activation and only supported and connected to the outside, through virtual reality. Even if the mechanical terminals incorporated an artificial skin, allowing the transmission of sensations to the brain, it would not have the ability to experience any sensation, without the chemical substrate present in the biological individual and developed over millions of years; would be matched to a colorless canvas, a lackluster and distant image, reflected in the cold glass of the temporal mirror.

Just as the perfect image of a solar system, it does not produce heat as López Corredoira of the Canary Islands Astrophysical Observatory has very aptly pointed out, a memory without emotion, incapacitates the complete experience; a rose would have more life, if it could be quantified. The memory of the first kiss, the exciting nervousness of a presentation at work or at university, the moments of evasion or the relaxing feeling of contemplating a calm sea, all of this, turned into a blurry, blurred and vague succession of images without emotion, a loss of vital wealth; the descent from the human level to the plant, achieving a long life time, clearly undesirable.

Artificial intelligence, even with millions of lines of computer code in its programming, would not be able to anticipate the enjoyment of the taste of a home-cooked meal by the aroma that comes out during its cooking; it could not form a mental map of something other than its circuits, nor experience the excitement of an impossible dream; faced with a paradox in which the imagination was specified, it would be blocked; I would never reach the solution to the challenge "imagine a knife without a handle or blade", so simple for a human. If it were to be called mechanical intelligence -a definition with more descriptive power- it would not have raised so many expectations. An incomplete body is the result of bionic promise, whether it comes down to the permanence of the mind, or to the rest of the body.

If life is experimentation, illusion, struggle and continuous tendency towards perfection, the virtual machine will never be able to live. Some would say it is not made in the image and likeness of God. The human being, endowed with a congenital imperfection, dreams precisely for the awareness of its limitation, with the scope of the ideal state. Fight, desire and permanent curiosity give you the most rewarding feeling possible: freedom. Only the human being has already become, without the need to wait for the future, the most perfect work of the known universe, by virtue of the most effective existing heritage, granted by Mother Nature.

Still connected to a global brain-net, composed of the network composed of all existing brains and endowed with unlimited learning capacity, an artificial intelligence, would be unable to produce the slightest abstraction or the actual generation of knowledge; nor would i experience post-progress gratification. Its fundamental usefulness is limited to question answering, even if the answer adopts a visual format loaded with "pixels", offering a three-dimensional high definition image, which is not little. Scientists should be asked not to give up their imagination, but not to fall, let alone propose, mere speculation. Playing with each other's emotions and desires equals the ancient shamans, healers and vaticinators of the future, that is, makes them practitioners of the ancient art of divination.

Post-humanism basically forms a great and novel narrative with background and civilizing implications. The human essence rests on the recognition of congenital imperfection and the irrational, irrepressible and unintelligible desire to achieve perfection, even by going to unrealistic scenarios, provided by its symbolic capacity. In this line, trans-humanism is another chapter, in the episodic narrative typically used in the descriptive format of human evolution. In the near future, it will be yet yet yet one more chapter of the historical episodic.

The current civilization, bombarded by the stalking dangers of modernity, such as climate change or poverty, is witnessing the explosion of a new myth of profound implications, a new banner to fly, a great goal to achieve, a supreme idea to adhere to. An alternative of enhancing potentials that overcomes simple progression, enabling an opportunity to knock on the gates of paradise, to allow the defense of a new cause of civilizing danger: the end of humanity. The transhuman movement, delays the limit of the known civilization, advocating the so often implored change of cycle, which already had other previous versions such as the beginning of the century or the intentionally debunked Maya prediction about the year 2012.

Philosophers, technicians, religious, politicians, all of them delighted with the emergence of the central issue that generates a new global debate. Debate, with clear moral, ethical and of course legal implications; a resurgent discursive element that allows the confrontation and resolution of chronic problems, determinants of inequality, poverty or indignity with the weak, to be circumvented; and above all, an ideological guideline, a new weapon for deniers and fans of the apocalypse.

The *technomito* has successfully met the shocking objective of global onslaught; has far surpassed the adoption of its multiple roles, as a great drain for civilizing drainage. A hope for the attainment of ascension to a human category of greater significance, to the excellent belonging of the legion of the children of a lesser God. At the same time, engulfed by business possibilities, it emanates the option of obtaining disproportionate economic benefit. In this line and within the prevailing neoliberal worldview, a new element generating oppression and also inequality. As an ideological element in its antipodes, the political promise of its use as a distinctive feature of equality and social progress. But above all, a call to the resurgence of human ancestral longing, condensed into denial of death, transgression of natural limits; a new technified path to impossible immortality, a side to hold on to for the prolongation of life, victory over physical barriers, the ultimate overcoming of time and matter; the definitive demonstration of the medieval anthropocentric ideal: the human being as the center and king of creation.

For those who like the exhibition, the large technological market offers fashion and avant-garde possibilities, a la cart prosthesis and connective and communication varieties; external additives for perfection. Again disregarding the principle defended by Eastern spirituality, based on introspection with a look into the being, where the true gods and human demons dwell; the real ground that must be conquered. Perhaps it was no coincidence that both spirituality and civilization were born in the East. Materialism and spirituality, constitute the basal assembly of the construction of both civilizations; enjoy and immediacy, in the face of containment and inner progress.

It is therefore not a question, as the illustrious and admired Michio Kaku has stated, to harbor the desire to be like God, but rather and fundamentally, to continue the path marked by the unalterable tendency that represents the ultimate expression of humanity: dreaming. The human being is never more human than when questioning, wanting, seeking, fighting, and even awake with all that without having seen and less being aware of its real existence, pursues and yearns; and most of the time, without the use of its maximum quality: reason. By definition, the dream is irrational but always legitimate and above all, free just as free is, the subject who dreams; the being that forever will be human.

Bibliographic Sources and References

Blogs

- Blogthinking.com. https://blogthinkbig.com/

Institutional Sources

- Amref Salud África. https://www.amref.es

- Bio-Cord. https://bio-cord.es

- Calicó. https://www.calicolabs.com/

- Humanity +. https://transhumanismo.org/pensamiento-y-opinion/

- Infovaticana. https://infovaticana.com

- Institute for Human Future. https://www.fhi.ox.ac.uk/

- La Nueva España. https://www.lne.es/

- La Red 3.0. https://www.redtrescero.es/es/

- MIT Technological Review. https://www-technologyreview-com.cdn.ampproject.org

- NCYT. https://noticiasdelaciencia.com/

- Fayerwayer. https://www.fayerwayer.com

- Open Mind. https://www.bbvaopenmind.com/ciencia/investigacion/

- Quanta Magazine: https://www.quantamagazine.org/

- Santa Fe Institute. https://www.santafe.edu/

- World Economic Forum. https://intelligence.weforum.org

Monographic Articles

- Acosta, M. (2019). "Inteligencia artificial: La cibernética del ser vivo y de la máquina. Naturaleza y Libertad. N.º 12. Revista CTS, n.º 41, págs. 287-311.

- Aracil, J. (2019). "El debate sobre la ingeniería y la ciencia". Revista CTS, nº 41, vol. 14, Junio de 2019 (pág. 287-311).

- Bateson, G; Bateson, M.C. (1994). El Temor de los Ángeles. Barcelona:Gedisa.

- Bruna Catalán,I. "Desde nuestro origen hasta nuestro futuro". Revista Iberoamericana de Fertilidad y Reproducción Humana ∕ Vol. 36 nº 2 Abril-Mayo-Junio 2019 ⁄3.

- Castañeda Bustamante, H. (2018) "La paradoja simplificadora del discurso del pensamiento complejo". En revista Encuentros, Vol. 16-02 de julio-dic.

- Castrejón González, E.O. "Simulación por Computadora: Una Herramienta Robusta". 13a Feria de posgrados de calidad. CONACYT. 11 de marzo de 2012.

- Durán, J.M. "Nociones de simulación computacional: simulaciones y modelos científicos". Universidad Nacional de Córdoba – Argentina CONICET – Argentina.

- Escudero, C. "Condición Poshumana y Otras Reflexiones Contemporáneas". Cátedra, (16), pp.95-107, agosto, 2019. ISSN 2415-2358,

- Elena Ortega, J.M. "La Singularidad Tecnológica: ¿Mito o nueva frontera de lo humano?". Naturaleza y Libertad. Número 12, 2019. ISSN: 2254-9668.

- Espíritu Ávila, A.R. "¿Es posible o no la inmortalidad cibernética?". Tierra Nuestra, 12: 41-51 (2018) ISSN 1997-6321 (Versión impresa) ISSN 1997-6496 (Versión electrónica) DOI: http://dx.doi.org/10.21704/rtn.v12i1.1268 © Universidad Nacional Agraria La Molina, Lima - Perú

- Frausto Gatica, O. "La Sociología de la Ciencia y la Reflexividad Científica". Acta Sociológica, núm. 67, Mayo-Agosto de 2015, pp. 193-220.

- García-Valdecasas Medina, J.I. "La simulación basada en agentes: una nueva forma de explorar los fenómenos sociales". Reis 136, octubre-diciembre 2011, pp. 91-110.

- Giménez, J.C. "Una Mirada a la Reproducción Celular". Revista Leds, núm. 11.

- Hernández García, I. "Estética de lo Imposible". DATJournal v.4 núm.2, 2019.

- Hipertextos, Capitalismo, Técnica y Sociedad en debate. Vol. 4-Número 6.

- Informe Quiral (2017) La Edición Genética ante la sociedad. Universidad Pompeu Fabra: Barcelona.

- López Corredoira, M. "Del Hombre-Máquina a la Máquina-Hombre: Materialismo, mecanicismo y transhumanismo". Naturaleza y Libertad. Número 12, 2019. ISSN: 2254-9668.

- López Medina, E, Mostacero León, J., Efraín Gil Rivero, A y De La Cruz Castillo, J. " Aproximación ontológica y epistemológica de la ingeniería genética". REBIOL 2018; 38(2): 66 - 76, Julio - Diciembre Revista Científica de la Facultad de Ciencias Biológicas. Universidad Nacional de Trujillo. Trujillo. Perú. ISSN: 2313-3171 (En Línea).

- Maldonado, C.E., Gómez Cruz, NA. "Modelamiento y simulación de sistemas complejos". Facultad de Administración. No. 66, ISSN: 0124-8219 Febrero de 2010.

- Parra Sáez, J. "Realidad y Ficción Transhumanista". Revista Laguna, 44; julio 2019, pp. 93-111; ISSN: e-2530-8351.

- Pedro Barrajón, L.C. "El futuro de la humanidad: perspectiva post-humanista". Ecclesia, XXXIII, núm. 1, 2019 - pp. 27-38

- Terrones, A. "Transhumanismo y ética de la responsabilidad". Revista Resonancias No. 4, 2018

- Terrones Rodríguez, A.L. "Una aproximación general al transhumanismo
 y su problematización". ANÁLISIS ISSN: 0120-8454 e- ISSN: 2145-9169 Vol. 51 / núm 95 Julio - diciembre de 2019 pp. 319-345.

- Vásquez Rocca, A. "Peter Sloterdijk: Normas para el Parque Humano, del Humanismo al Posthumanismo". Encuentros Multidisciplinares, nº 62 Mayo-Agosto 2019.

Printed Books

- Cammilleri, R. (1994). Los Monstruos de la Razón. Madrid: Ediciones Rialp.

- Castells, M. (2004). The Corneto Society: A Cross-Cultural Perspective. Northampton, M,A: Edward Elgar.

- Feyerabend, P. (1996). Adiós a la razón. Madrid:Tecnos.

- Kaku, M. (2014). El Futuro de Nuestra Mente. Barcelona:Debate.

- Kuhn, Th. (2000). La Estructura de las Revoluciones Científicas. México: Fondo de Cultura Económica.

- Morín, E. (2006). El Método. Madrid: Debate.

- Touraine, E. (2006). Crítica de la Modernidad. México: Fondo de Cultura Editorial.

www.ingramcontent.com/pod-product-compliance
Lightning Source LLC
Chambersburg PA
CBHW070619220526

45466CB00001B/54